Growing Grapes in Texas

Texas A&M AgriLife Research and Extension Service Series

PATRICK J. STOVER AND JEFF HYDE, *General Editors*

Growing Grapes in Texas

*From the
Commercial
Vineyard*

*to the
Backyard
Vine*

JIM KAMAS

TEXAS A&M UNIVERSITY PRESS

College Station, Texas

Manufactured in China through Martin Book Management
This paper meets the requirements of ANSI/NISO Z39.48–1992 (Permanence of Paper).
Binding materials have been chosen for durability.

Library of Congress Cataloging-in-Publication Data

Kamas, Jim, 1955– author.
Growing grapes in Texas : from the commercial vineyard to the backyard vine /
Jim Kamas.—First edition.
pages cm—(Texas A&M AgriLife Research and Extension Service series)
Includes index.
ISBN 978–1-62349–180–2 (flexbound : alk. paper)—
ISBN 978–1-62349–223–6 (e-book)
1. Viticulture—Texas. 2. Vineyards—Texas. I. Title.
II. Series: Texas A&M AgriLife Research and Extension Service series.
SB387.76.T4K36   2014
634.809764—dc23
2014015245

Unless otherwise credited, all photographs are from the author's personal collection.

To my family—
my loving wife Jacy,
my grown kids Nicolas and Rebecca,
and our new addition, baby Jonah.
You all make my life a joy.

Contents

Foreword, *by Larry A. Stein*   ix
Preface   xi

1. Why Do You Want to Plant a Vineyard?   1
2. History of Grape Growing in Texas   7
3. Grape-Growing Regions of Texas   12
4. Factors Limiting Grape Production in Texas   19
5. Systematics of the Genus *Vitis*   35
6. Choosing a Vineyard Site   44
7. Time line for Establishing and Planting the Vineyard   53
8. Vineyard Design   65
9. The Growth Cycle and Grape Maturity   70
10. Grapevine Physiology   82
11. Rootstock Selection   93
12. Selecting Fruiting Varieties   104
13. Choosing a Training System   115
14. Pruning and Training Dormant Vines   123
15. Canopy Management   139
16. Grapevine Nutrition and Vineyard Fertilization   149
17. Diseases Affecting Foliage and Fruit   167
18. Other Diseases Affecting Grapevines   184
19. Vineyard Floor Management   191
20. Grapevine Water Needs and Vineyard Irrigation   202
21. Insects and Mite Pests   210
22. Vertebrate Pest Control   221
23. Vineyard Equipment and Infrastructure Needs   225
24. Vineyard and Winery Relations   235

Glossary   241
Index   245

Foreword

What an honor to be able to provide a preview of what will become a "must have" book for every viticulturist in the state, from the beginning grape grower to that experienced grower looking for answers to everyday problems. Jim draws on his broad experiences to answer questions that potential or current growers did not even know they needed to ask. The book is not only comprehensive but realistic and easy to understand, though very technically accurate. Jim not only shares his philosophy but carries you through all the vital steps needed to be successful in growing grapes. From history, to grape regions in Texas, to rootstocks and varieties, to training and pruning, to controlling diseases and insects—it is all here. He draws on his own trials and tribulations to help others avoid common pitfalls. Along the way he offers commonsense tips on the hows and whys of what it takes to be a successful grape grower serving the wineries of Texas.

I have known Jim since our days as graduate students, and I can assure you that he lives and breathes perennial fruit crops in Texas and has always had that special passion for grapes. He thrives on solving the nagging problems that seem to crop up from time to time. So hang on for not only an exciting but a worthwhile read that will definitely either inspire you to want to be a grape grower or help you overcome the many challenges of growing grapes in Texas.

—LARRY A. STEIN
Texas A&M AgriLife Extension Service

Preface

Early in my college career I fell in love first with plants, then later specifically with perennial fruit crops. I loved working in a fruit breeding and variety development project at Texas A&M, and later I was quite content to be a commercial peach grower in Austin County, Texas. Nature and fate had other intentions for me, and after I endured three complete freeze-outs in my orchard, I was given the greatest professional opportunity of my life. In 1988, I was offered, and I accepted, a job as Cornell University's extension viticulturist in western New York. My first day on the job, May 9, we had a five-inch snowfall. The first cold front blew in on August 15. I was obviously not in Central Texas anymore. Although I was in culture shock, to my delight I found myself living in a progressive community whose economic viability was centrally focused around grape production. These people had been growing grapes for generations and were among the most warm, receptive people I have ever known. They adopted me, were patient with me, and gave me more than I could ever give back. To this day I have many dear friends in western New York that mean the world to me.

I worked at Cornell's Vineyard Lab in Fredonia, on the shore of Lake Erie, and spent the next eight years learning all I could absorb from the incredibly competent and knowledgeable people who surrounded me. Rick Dunst, my research counterpart at the Fredonia lab, quickly became, and remains, my best friend. He and I still talk frequently, we visit each other in New York and Texas as often as we can, and we share the same passion for field viticulture. Bob Pool, Roger Pearson, Wayne Wilcox, T. J. Dennehy, Martin Goffinet, Alan Lakso, Jerry White, Wes Gunkle, and countless others were, and in some cases still are, incredible resources, colleagues, and teachers who had a tremendous, positive impact on me.

One of my greatest teachers was Dr. Nelson Shaulis, arguably one of the best viticultural minds in the world during the twentieth century. I started working for Cornell after Nelson had retired, but there was never a time he was not active in viticulture. I remember being horribly intimidated by him when we first met, and he continued to challenge me at every opportunity for the eight years I lived and worked in New York. I worked hard to try to reach his standards, and I came to understand that his actions and words were simply those of a great teacher and a great mind. His attention to detail, in viticulture and in the written and spoken word, has had a profound effect on me both professionally and personally. He provoked us all to think and to question our own assumptions. On more than one occasion he is known to have challenged his audience by asking the question, "Tell me, which is the greater impediment to the advancement of one's viticultural knowledge: failing to answer questions, or failing to question answers?"

With that, I encourage each of you to keep that open mind Nelson called us all to maintain. Seek the counsel and advice of many others, but choose the advice you follow carefully. Proceed cautiously and always question not just the "how" but the "why" of your decisions. I sincerely hope that the chapters that follow will help you navigate the endeavor you have chosen to begin.

Finally, thanks to the entire team at Texas A&M University Press, especially its editor-in-chief, Shannon Davis, for her support and guidance throughout this project; copyeditor Maureen Bemko, and project manager, Patricia Clabaugh, who shepherded this book through the publications process. Thanks also to Jacy Lewis for her first-rate work on the index.

Growing Grapes in Texas

Why Do You Want to Plant a Vineyard?

ALTHOUGH IT MAY SEEM somewhat rudimentary, it is extremely important for persons or families who want to get into the vineyard business to ask themselves what is behind their desire. Is it because of the pleasure in drinking wine? Is it the social aspects of winery events? Is it the pleasure of working outside and growing things? Grape growing is a venture that requires constant infusions of time, energy, and resources, and, like dancing, timing is everything. A fungicide is the same price if purchased and applied before or after an infection period, but the outcomes will be entirely different.

The most common mistake new and prospective growers make is to

Planting and tending a vineyard can be a rewarding experience, but it takes far more time and resources than most prospective growers realize.

underestimate how much time it takes to successfully establish and manage a vineyard of any appreciable size. So, in the beginning, ask yourselves these questions: Are we ready to devote *every* weekend to this venture? Are we willing to spend *all* of our vacation time pruning, harvesting, and managing a vineyard? Many couples entering the business find that, having purchased a vineyard, their "together" time is spent *solely* erecting trellis, tying up young vines, and spraying weeds. This scenario may sound like an exaggeration intended to discourage everyone from even considering establishing a vineyard. That is not the case, but it is a call to carefully examine all aspects of the enterprise before you get involved in it. The impact on lives and relationships is real. Owning a vineyard can be a uniquely gratifying experience that is professionally and emotionally rewarding, or it can end in heartbreak and economic perdition. It's a sad commentary, but I have a computer file of photographs of abandoned vineyards. The file name is "Broken Dreams." While the narratives of how these vineyards ultimately failed are tragic, far worse is the tale of how their demise affected many of those families. While successful grape growing is founded on scientific principles, there is still a strong element of risk. As with any temptation of fortune, people should never gamble any more than they can afford to lose, financially or emotionally.

Unfortunately, this scene reflects the broken dreams of many an aspiring grape grower who became overwhelmed by the demands of the task.

The second greatest misconception among prospective growers is the notion that a vineyard is a simple thing to manage. Fully understanding grapevine physiology, light interception, response to soils and climatic variation, and fruit maturation, as well as the abundance of diseases and insect pests, fungicide modes of action, and herbicide chemistry, to name a few important concepts, are all critical to making sound management decisions vital to the health and productivity of a vineyard. Most prospective growers tend to come from the ranks of the successful. However, being a skilled neurosurgeon or attorney does not automatically make someone a qualified vineyard manager. If owning a vineyard is one's dream, even the supremely intelligent and hardworking professional must prepare for an entirely new course of study. A successful grape grower must know not just the "hows" but the "whys" of making management decisions. Most economically successful grape growers were successful first in some other commercial agricultural enterprise, such as cotton, peanuts, or vegetable crops. Find such persons and seek their advice; they have survived for a reason.

The biggest mistake commonly made by new growers is to plant a vineyard on a parcel of land because they already own it. They have looked over a field and remarked how beautiful that land would be with a vineyard on it. If you want to plant a small vineyard for noncommercial grape production, simply to supply your friends and family with a source of grapes for home wine production, then establishing a vineyard because you already own the land makes sense. However, if the goal is to create an economically viable commercial grape vineyard, using land simply because you own it may be a mistake. Site selection is the single most economically critical decision a prospective grower can make. Choose the land for your venture only after careful consideration, and with abundant caution. Grape growers have an adage that speaks to this principle: "Live where you farm, don't farm where you live." While it's important to live in proximity to your vineyard, it's wiser to move your house close to your vineyard than to choose a vineyard site simply because it is near your current homestead.

Challenges to Economic Gain

With prices for some grapes exceeding two thousand dollars a ton in 2013, a vineyard might appear to offer its owner the potential to make vast sums of money. This view is generally deceptive because of the high financial inputs required to ripen a crop of high-quality fruit. Grape growers and wineries alike share the old adage that to make a small fortune in the business, one must first start with a large one.

While many factors need to be considered to make a vineyard a viticultural success, one element is common to all enterprises that are an economic success: a realistic business plan. One of the biggest challenges facing a new grower is that of economic scale. In other words, it takes the same tractor, the same airblast sprayer, the same mower, and so forth, to care for a two-acre vineyard as a twenty-acre vineyard. These fixed economic costs present a greater economic challenge to a small operation that will have lower returns than a larger one. Likewise, new growers cannot expect to command premium prices for their fruit until they have proven to a winery that they can consistently predict crop size and deliver superior quality fruit. Successful growers maintain long-term contracts with one or more wineries and develop a working relationship that features a win/win attitude. Not all wineries have this attitude, and not all growers can manage their vineyards to give sustainable yields of superior quality fruit. It is true that, in 2013, Texas wineries are in great need of additional fruit. This does not mean that anything you grow will be purchased, however. One can make bad wine out of good grapes, but one cannot make good wine out of bad grapes. The current market is very open and competitive, and, over time, wineries that make bad wine will not survive.

Labor Issues

Those who have a desire to run a vineyard because they like to do hands-on work outdoors and to grow things should recognize the reality of that dream: killing weeds in August when the mercury climbs to 110 degrees Fahrenheit, pruning vines when it's freezing cold because you absolutely have to start when it's still winter to finish the job by spring, and dodging black widow spiders while you harvest. These are the realities of grape growing. While you and your family may be able to endure these extremes, the reality is that, to be economically viable, a vineyard operation needs to be large enough that your family can't do everything themselves.

Finding additional workers who will endure these harsh conditions is one of the most daunting challenges of owning a vineyard. Although the United States entered a prolonged period of high unemployment some years ago, most advertisements for seasonal farm labor go unanswered or applicants quickly leave after the realities of the situation reveal themselves. Any discussions of farm labor involve the contentious politics of immigration reform, but most who are actively involved in commercial agriculture will agree that some form of a legal, simplified guest worker program is sorely needed. The current farm guest worker program, known as the H-2A program, is a bureaucratic nightmare. While some fruit and vegetable growers still use workers with these H-2A-type visas, the problem with the program is that many federal lawmakers do not believe that American agriculture actually needs additional workers. Herein lies the dilemma. To be efficient and economically viable, growers need to have an operation that is too large for a family to handle themselves, so they must either hire workers (which is extremely difficult) or work toward mechanization of vineyard tasks, which requires expensive specialized equipment. Once again, the economies of scale become a critical aspect of the enterprise.

Accepting Risk

No matter how carefully a site is chosen or varieties and rootstocks selected, grape growers are constantly challenged by different elements that throw risk into the equation. Weather across Texas, and indeed most of the southern United States, is quite variable, and extremes in weather conditions constantly affect viticultural management decisions. Ed Auler, longtime winery owner of Fall Creek Winery in Tow, Texas, describes viticulture in Texas as "growing grapes under periods of extended drought only to be interrupted by an occasional flood." His observations are indeed indicative that weather averages are made up of extremes. There are few "normal" years, so having a site whose features can help moderate these dramatic shifts in conditions is necessary for long-term profitability. Not only rainfall but temperature extremes complicate consistent grape production. Freeze injury caused by winter low temperatures can cause bud, cordon, and trunk injury or death. Spring frost can destroy a crop in a few hours, as can hail, and heavy rainfall near harvest can lead to rotten fruit. There are a few cultural practices growers can employ to lessen the effects of weather extremes, but site selection remains the single most critical factor affecting outcomes.

So, for those still interested in reading further and jumping into this venture, here are a few adages to bear in mind:

PLANTING A VINEYARD IS FAR MORE WORK THAN YOU EVER DREAMED.

- There is much more to learn about growing grapes than you can imagine.
- Establishing a vineyard will cost you more than you planned on spending.
- Remember Murphy's Law: "Whatever can go wrong, will go wrong."
- And don't forget O'Leary's Corollary: "Murphy was an optimist."

History of Grape Growing in Texas

THE HISTORY of grape growing in Texas predates that of California by nearly a century. In the 1660s, Franciscan monks brought grape cuttings from Mexico and established vineyards for sacramental wine production at the mission at Ysleta on the Rio Grande, near present-day El Paso. The success of these vineyards was probably due to the disease tolerance and easy adaptation to local conditions of the Mexican nursery stock, and those vineyards remained economically viable until the early twentieth century. Historical accounts indicate that the wave of European settlers from wine-producing countries in the mid- to late nineteenth century brought *Vitis vinifera* grape cuttings from the Old World, and there are records of attempts to establish vineyards near Bell-ville, New Braunfels, and Fredericksburg. There are no reports of notable production from these vineyards, and by all accounts, they soon failed. These settlers soon learned that by adding sugar to the juice of several wild, native Texas grape species, stable wine could be produced.

Gilbert Onderdonk (1829–1920) began working with horticultural crops at an early age, and, by the time he was eleven, he had developed several new varieties of Irish potato. Onderdonk moved from his native New York to Texas in 1851 for health reasons, and he eventually bought 360 acres of land in Victoria County, Texas, where he taught school, raised horses, and, about 1858, started a fruit nursery. Gilbert Onderdonk volunteered and served in the Eighth Texas Infantry, and he participated in battles at Corpus Christi and Fort Esperanza during the Civil War. In subsequent years, he was regarded as one of the leading horticulturists of his time, and at his Mission Valley Nursery he experimented with, propagated, and sold grapevines from the Rio Grande to New Orleans, all of them well adapted to local conditions.

Fellow horticultural pioneer Thomas Volney Munson considered him an inspirational contemporary, and Onderdonk in turn recognized the importance of Munson's work. Onderdonk undoubtedly encountered the grapevine malady known as Pierce's disease (PD) and was among the first to recognize the "blight" we now know as cotton root rot. In his writings he noted, "We had temporary success with *labrusca* and some *vinifera* varieties. But, from different causes, we found them unreliable and short lived with us." He further stated that "in most occupied portions of Texas, there are spots of ground upon which cotton and some other plants die out . . . all of the grasses seem unaffected by it. This blight is quite sure to kill every apple and pear tree and every grape vine which it attacks and sometimes destroys peach trees and rose bushes. I have seen a whole orchard—one by one, in regular succession—yield to its withering power."

Munson (1843–1913) was born in Illinois, graduated from Kentucky State Agricultural College, and moved to Denison, Texas, in 1876. Munson apparently developed his love of breeding grapevines from visits to his chemistry professor's vineyard in Kentucky during the fall of 1873. Although he, like his brothers, was involved in real estate businesses in Denison, Munson quickly found himself captivated by the tremendous biodiversity he found in Texas. From 1880 to 1910, he collected, catalogued, and bred grapes from native southern species and developed more than three hundred new grape cultivars adapted to various areas of Texas and the southeastern United States. In addition to being cold hardy and resistant to fungal pathogens, many of Munson's varieties were tolerant to Pierce's disease. In 1909, Munson published *Foundations of American Grape Culture,* which became the standard text on grapevine genetics and culture in its day. Munson's greatest fame, however, came from his recognition that grape phylloxera, an insect that was decimating vineyards across Europe and indeed most of the world, could be overcome by grafting susceptible vines onto native Texas grapevines, which were resistant to the pest. For his efforts, the French government sent a delegation to Denison to bestow the French Legion of Honor, Chevalier du Mérite Agricol, on Munson. He received numerous other honors and awards, and in villages across France there are statues and plaques acknowledging the work of T. V. Munson.

In 1883, an Italian immigrant named Frank Qualia established Val Verde Winery in Del Rio, to grow 'Mission' grapes for wine production. Around 1890 vines started dying (most likely due to Pierce's disease), and by 1910 the vineyard was replanted with 'Lenoir,' 'Herbemont,' and

'Champanel' grapevines. Now more than a century old, Val Verde Winery is the only winery in Texas that survived Prohibition, and the Qualia family still successfully operates the winery today. Val Verde Winery was among the first to show that long-term survival and success is possible even for vineyards in that part of the state.

Around 1900 agricultural reports and bulletins showed interest in experimental grape plantings across the state. By 1900 Texas had more than twenty-five wineries, but Prohibition brought an end to industry expansion. Munson established a grape nursery and sold nursery stock, with the profits going to further his grape exploration and breeding efforts. When the Munson & Sons Nursery closed in Denison, the collection was moved to the Winter Garden Experiment Station at Winter Haven, Texas. These varieties became part of the extensive grape evaluations that Ernest Mortensen began in 1931 and were terminated when the station closed in 1952. Grape evaluations were also conducted from 1939 to 1963 by Uriel A. Randolph at the experiment station near Montague. In addition to doing variety evaluations, the station staff conducted fertilization, pest management, and rootstock trials.

From the late 1960s through the 1970s Texas experienced a resurgence in grape growing. Seeking higher wine quality, growers switched their variety selection from American varieties to French-American hybrids. In 1974, Ron Perry published Texas Agricultural Experiment Station Report 74–3, entitled *A Feasibility Study for Grape Production in Texas.* In that study, Perry identified Pierce's disease as the number-one limiting factor in Texas grape production, and he produced a map outlining the probability of Pierce's disease incidence across the state. At that point there was a rudimentary understanding that the distribution of the pathogen was limited by cold winter temperatures. It was thought that disease development was limited to areas receiving less than eight hundred hours of winter chilling per year. It was also thought that the range of vectors was limited to humid areas of the state. Perry's study also cited cotton root rot, winter injury, spring frost, and hail damage as additional limitations that would curtail further growth and prosperity in the Texas wine grape industry.

By the 1970s grape planting had once again increased in several regions of Texas, but much of the acreage was planted to French-American hybrid varieties because of their cold hardiness and disease resistance. Wine quality from these varieties left much to be desired, and wine sales were largely driven by regional pride. Against the advice of many experts,

Two dedicated professionals who ushered in the new age of Texas viticulture in the late twentieth century are:
▶ Bill Lipe, a grape researcher at the Texas Agricultural Experiment Station at Lubbock, and
▼ George Ray McEachern, a grape extension specialist at Texas A&M University in College Station.

growers began to plant some *Vitis vinifera* varieties in the High Plains, Hill Country, and North Texas. Just as people buy fresh fruit based on cosmetic appearance, wine sales are commonly driven by varietal name recognition by consumers. Chardonnay, Merlot, Cabernet Sauvignon, Sauvignon Blanc, Riesling, Pinot Noir, and Gewürztraminer wines began making their way to the shelves, and wine quality was on the rise. As more winemaking talent found its way to, or back to Texas, local wines began being recognized for their quality. The introduction of 'Rhône,' 'Port,' and other hot weather varietals, much more adapted to ripening in the heat of Texas summers, began to be planted, and once again, wine quality increased dramatically.

Today Texas winemakers and their wines have clearly established that their products can compete with the best wines of the world. While winery numbers continue to climb, planted grape acreage lags behind the demand created by this booming market. Many, many giants in the industry, from academic viticulturists and grape-growing pioneers to talented winemakers, have dedicated their careers and indeed their lives to helping this industry grow. Other writers have eloquently chronicled the lives of these forerunners, and the whole industry is deeply in their debt. With a new generation of visionaries at the ready and with obstacles to profitable, sustainable grape production being overcome, the future of the Texas wine industry appears bright.

Grape-Growing Regions of Texas

EXAS HAS six distinct regions where conditions differ enough to affect the growing of grapes. The details offered for each region are merely summaries; potential growers should seek more in-depth information on regions of particular interest.

High Plains

The High Plains, as this region is known in Texas, or the South Plains, as the area is known nationally, is the largest commercial grape production region of the state. Affordable farmland with deep soils and an ample water supply make the High Plains a logical choice for many grape-farming options. Another reason this area is attractive for planting vineyards is that it already has people engaged in farming of various kinds. For many vineyard owners, farming was their first and, in some cases, their only occupation. While growing grapes is very different than growing an annual crop such as cotton, having land, a farming mind-set, and at least some equipment is a good starting point for a new High Plains grape grower.

One of the advantages of growing grapes on the plains is the reduced risk of some grape diseases. While the bacterial infection known as Pierce's disease, or PD, is widespread among vineyards, it appears to be a chronic problem rather than an acute one. It has yet to be isolated from grapevines, and it appears that many infections may be held in check by cold winter temperatures. Many fungal diseases, while present, appear to be less problematic than in more humid portions of the state. Powdery mildew remains the primary fungal pest growers need to manage.

As with any area, there are risks to growing grapes on the High Plains. Over the years, extreme winter freezes, either in late fall or following a winter warming spell, have led to crop loss and severe vine injury. Other

A vineyard on the High Plains.

parts of the state certainly have this risk as well, but the chances of devastating freezes are higher on the High Plains. This region also has a growing season that features frequent hailstorms, which can not only harm or destroy the crop but also seriously damage the vines.

Another problem is that the region features many cotton fields where 2,4-D herbicides are used for pre-plant weed control. Grapevines are extremely sensitive to phenoxy herbicides like 2,4-D and may suffer serious damage from overspray. While the parties applying the herbicides are responsible for any damage these off-target drifts cause, pinpointing who sprayed what and when can be mind-boggling, especially when so much herbicide is being sprayed. All of these risks need to be weighed along with the advantages of grape production on the High Plains.

Trans-Pecos

The Trans-Pecos region is home to the state's largest vineyard, Sainte Genevieve, and has numerous vineyards, from Fort Stockton to Marfa, Fort Davis, and beyond. The higher altitude offers cooler temperatures during grape ripening, and this region has produced some outstanding wines. The limitations of the Trans-Pecos region are a questionable supply of high-quality water and the tremendous temperature swings during winter, which can cause devastating freeze injury in the vineyard. The allure of this area's natural beauty may be reason enough to consider establishing a vineyard in the Trans-Pecos, but, as was true a century ago, it takes a rugged soul to survive the extremes that come with this territory.

Cross Timbers and Red River Valley

The gentle rolling hills of the West Cross Timbers and the area south of the Red River in Texas have many things in common. Winters are colder than areas to the south, a differential that has both negative and positive attributes concerning growing grapes. We now know that the range of Pierce's disease clearly includes these areas, and although this disease can kill vines, colder temperatures probably mitigate the lethality of the disease to some degree. Of course, cold temperatures may also injure vines and cause freeze/frost injury to the crop. That said, there are many fine vineyards in this part of the state, and they play an important role in supplying the many wineries in the Dallas/Fort Worth area with the grapes they need.

Prospective growers wishing to plant in this area need to consult topographical and soil maps carefully and then choose elevated sites to avoid areas prone to winter injury. While much of the soil in this area is mildly

Vines growing in the Trans-Pecos region.

alkaline, there are pockets of neutral or slightly acidic soils that are pre-ferred for nutritional reasons and for soil-borne pathogen (cotton root rot) resistance. Another site selection criterion that should be carefully investigated is water quality. Within this region, especially in some areas along the Red River, the groundwater has very high sodium levels. Water testing should be conducted to determine sodium levels and the sodium absorption ratio (SAR). Potential sites with water SAR levels above 7 should be avoided.

The rolling hills of the region mean that site selection plays a large part in reducing risk from winter freezes and spring frosts. Hillside vineyard sites that allow for air drainage and also have adequate soil depth are far more common in this region than in many parts of the state to the south. One of the big advantages of grape growing in this region is the ready market provided by numerous wineries in or near the Dallas/Fort Worth metroplex. High-quality grapes are in short supply in Texas, and many wineries in this region provide the opportunity for growers to develop long-term relationships, thus providing both a measure of financial stability for the grower and high-quality fruit for the winery.

Texas Hill Country

At present, the Texas Hill Country is the second most visited wine destination in the United States, behind Napa, California. With more than three million people within a one-hundred-mile radius and abundant natural beauty, the Hill Country is a natural tourist destination. That said, the Hill Country can be a harsh place to grow grapes. The caliche hills, escarpments, and alluvial valleys define the dilemma of the Hill Country: where there is soil, there is no air drainage, and where there is air drainage, there is no soil. There are of course exceptions, but within this area, most of the locations with deeper soils are in low-lying valleys or creek bottoms that are prone to frost and freeze events. These alluvial soil deposits produced by rivers and streams also appear to bring with them increased risk of cotton root rot. The higher areas with good cold protection typically have thin, rocky soils. Finding the anomalies that defy this standard is the key to finding good vineyard sites in the Hill Country.

While typically less humid than areas east of the Balcones Escarpment, the Hill Country can be a land of extremes when it comes to rainfall. Extended droughts can be interrupted by tropical weather systems that deliver average annual rainfall in one week, causing sites to be saturated if not flooded. Because extremes of rainfall and drought are the rule

Grape growing in the Hill Country.

rather than the exception, new growers must find sites with ample water. Water supply may be dictated not only by the presence of water resources but also local groundwater management district regulations; some districts limit water production from local wells. In addition, new growers face the challenge of finding sites with both good surface and internal water drainage that will minimize the effects of tropical rain events.

South Texas

South Texas is a growing region characterized by a long, hot growing season, mild winters, and big fluctuations in water availability. Marginally to truly subtropical, the climate of the Lower Rio Grande Valley and counties to the north is capable of ripening crops in early June, before hurricane season, thereby avoiding the problems associated with high rainfall immediately preceding harvest. These characteristics are not assurances, however, and tropical rainfall still poses perhaps the greatest weather threat to the production of high-quality wine grapes from this region. There is also local interest in growing table grapes, and if the problems of bird predation, available labor, and visible fungicide residue can be overcome, fresh market table grapes can be produced very early in the season compared to other US production regions.

Perhaps the greatest challenges of growing grapes in South Texas are disease and the chemical properties of the soils in this region. Because Pierce's disease pressure is very high, 'Black Spanish' and 'Blanc DuBois' are the varieties most commonly grown in this area. High soil pH coupled with tight, silty clay soils limit the ability of these own-rooted, or self-rooted, varieties to extract sufficient iron to remain green and photosynthetically efficient. In addition to causing nutrition problems, soils in South Texas commonly exert tremendous cotton root rot pressure. High temperatures and tight alkaline soils create conditions that favor this pathogen, thus generating far greater disease pressure than encountered in the Hill Country. Plans for significant peach production in South Texas were also dashed in the late 1970s and 1980s by these same two limitations. At some point, the development and use of appropriate grape rootstocks will allow for continued expansion of grape acreage in the Lower Rio Grande Valley.

East and Southeast Texas

East Texas is generally characterized by deep, sandy soils with neutral to acidic pH and abundant sources of high-quality irrigation water. Good soils and water suggest that one can grow anything in East Texas. Unlike alkaline soils, the pH of which is virtually impossible to adjust in the short or medium term, acid soils can be limed to bring the pH closer to neutral, thus avoiding the aluminum toxicity common to many grape varieties grown on soils with a pH below 5.5.

For most of East Texas, Pierce's disease pressure is extremely high and not easily overcome by the cultural practices that have proven effective in the Hill Country. While there are a few brave souls cultivating susceptible *vinifera* varieties, most of the grape production in this region currently is limited to 'Blanc DuBois,' 'Black Spanish,' and small acreages of 'Favorite' and 'Champanel.' As more PD resistant/tolerant varieties are released, more choices will be available for growers and wineries to consider for commercial plantings. This region has perhaps its greatest growth yet ahead of it.

East Texas is blessed with higher rainfall than much of the rest of the state, but for grape growers that abundance means increased fungal disease pressure and the problems associated with achieving optimal fruit maturity under difficult conditions. Present and future variety choices should be made with disease resistance in mind, while canopy management and highly effective disease control programs are vital for overcoming this limitation of the region.

An East Texas vineyard.

Growers and wineries in East Texas will also need to develop long-term business strategies that are based on shared risk. Growers must strive to produce the highest quality fruit they possibly can, but wineries must also understand that "letting fruit hang" under high rainfall conditions will not necessarily improve fruit quality and that there are always yield losses when that practice is employed. For wine production in this region to be sustainable, both grower and winery must achieve success.

CHAPTER 4

Factors Limiting Grape Production in Texas

G IVEN THE VAST SIZE of Texas and its great variety of soils and climate zones, there are but three primary factors limiting grape production in the state. Each of them presents a host of issues to be addressed.

Pierce's Disease

Pierce's disease (PD) is a bacterial disease of grapevines that is ultimately fatal to susceptible varieties. Susceptibility means that, to varying degrees, a specific variety or vine does not have the capability of resisting increased colonization by a pathogen and begins to exhibit progressive symptoms until the vine ultimately dies.

The bacterium (*Xylella fastidiosa*) that causes Pierce's disease enters the vine through insect feeding and colonizes the xylem, or water conductive vessels, where it congregates, forms biofilms, and triggers organelle swelling within the vine, all of which ultimately disrupt the flow of water through the vine. The disease is vectored, or transmitted, by numerous xylem-feeding insects, which acquire the bacterium from other plant species near the vineyard or from infected vines within the vineyard.

The disease is native to Texas, and native grapevine species have evolved to be tolerant of the disease. Tolerance means that the vines can become infected yet also survive and remain healthy and productive because they have internal mechanisms that suppress bacterial numbers or limit the organism's mobility throughout the vine.

Determining the Range of Pierce's Disease

In 1974 Ron Perry identified Pierce's disease as the number-one factor limiting the production of grapes in Texas, and his report included a map outlining the probability of disease incidence across the state.

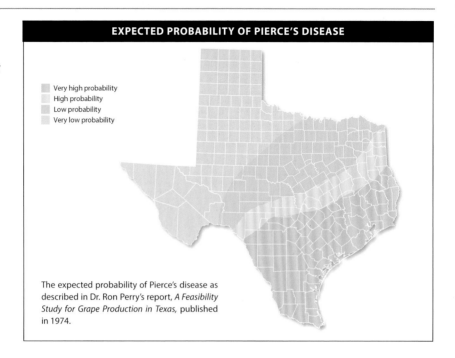

EXPECTED PROBABILITY OF PIERCE'S DISEASE

Very high probability
High probability
Low probability
Very low probability

The expected probability of Pierce's disease as described in Dr. Ron Perry's report, *A Feasibility Study for Grape Production in Texas,* published in 1974.

At the time Perry issued his report, there was a rudimentary understanding that the distribution of the pathogen was limited by cold winter temperatures. It was thought that disease development occurred only in areas receiving less than eight hundred hours of winter chilling per year. It was also thought that the vectors' ranges were only in humid areas of the state. Management guidelines prior to 2005 were based on research in California, where the disease did not appear until the 1880s. Only after a decade of applied research did the true dynamics of Pierce's disease epidemiology in Texas become apparent.

We now know that cold temperatures play an important role, directly affecting the ability of a vine to resist continued colonization after it has been infected with the PD causal agent. It is theorized that when vines are exposed to cold air, they produce cold-shock proteins that directly attack the bacterium within the vine and generate a curative effect. This theory explains that, although PD has been found as far north as Missouri, Tennessee, and northern Virginia, the destructive economic consequences of the disease are limited to areas that typically experience milder winters or a series of abnormally warm winter seasons. Between 2002 and 2012, insect and disease survey work has revealed that Pierce's disease exists in every grape-growing area of Texas, although the risks from economic loss

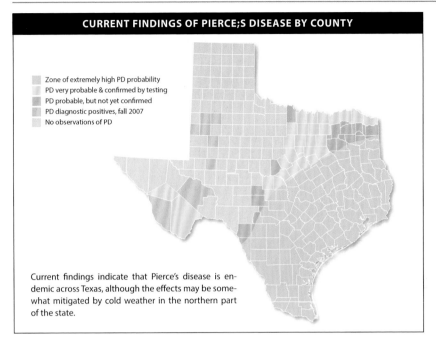

CURRENT FINDINGS OF PIERCE;S DISEASE BY COUNTY

Zone of extremely high PD probability
PD very probable & confirmed by testing
PD probable, but not yet confirmed
PD diagnostic positives, fall 2007
No observations of PD

Current findings indicate that Pierce's disease is endemic across Texas, although the effects may be somewhat mitigated by cold weather in the northern part of the state.

vary greatly by region. Deep southeastern Texas and the entire Gulf Coast region are at extremely high risk of vine loss due to the abundance of supplemental sources of the bacterium and very high insect vector diversity and population numbers. In short, grapevines in this region face challenge by the pathogen many times a day throughout most of the growing season. Such high disease pressure makes growing susceptible varieties incredibly risky and difficult in this part of the state. Even with the most rigorous management, infection is likely and vine loss will be high in this region. Choosing varieties that are resistant or tolerant will most likely continue to be the recommendation for very high-risk areas. Outside of this "hot zone" of extremely high PD probability, management practices such as site selection, floor management, the use of neonicotenoid insecticides, and rouging, or removal, of infected vines appear to have greatly mitigated the risk of vine loss due to Pierce's disease. Throughout the Texas Hill Country and the West Cross Timbers area of Central Texas, PD no longer appears to be the lethal threat it was prior to 2000, but areas of very high risk still exist within these geographic regions. Growers should be wary, however, because climatological conditions (cold winters and extended drought) most likely suppressed both vector numbers and pathogen survival in winter conditions. It is predicted that normal

In the late 1990s an outbreak of Pierce's disease caused widespread vine losses in parts of the Texas Hill Country not thought to be at high risk.

swings in temperature and rainfall will result in higher disease pressure over time.

The biggest surprise from research between 2002 and 2012 is the confirmation of PD not only in far West Texas but also across the High Plains of Texas. The presence of the PD causal organism has been confirmed in every grape-growing area of Texas, but the consequences of infection appear to be quite different. Although *Xylella* appears to be very widely spread among most of the vineyards in the High Plains region, the disease appears to be mainly a chronic stress rather than an acute one causing vine death.

It is postulated that while the bacterium may have been established in this region for quite some time, human activity such as the planting of cuttings from infected vines or the importation of infected nursery stock probably played a role in the widespread establishment of the disease. While there have been no documented cases of vines dying from PD on the High Plains, the same cannot be said for far West Texas. Grape vineyards near Fort Davis, Marfa, and Alpine have all suffered vine loss from PD even though the winter conditions in those regions were once considered sufficient to cure infection. The insect vectors in both of these regions are quite different from those in Central and East Texas, but they are abundant and quite capable of moving the disease.

Effective management of risk for Pierce's disease involves an eleven-step program of action:

1. Determine Relative Risk. When seeking prospective vineyard locations in Texas, you can get a good idea of the relative disease pressure by using historical maps. Do not deceive yourself into believing that, just because you own or are very attached to a particular field or an area, the risk of Pierce's disease will go away with willpower. For growers outside of Texas where Pierce's disease exists, areas that receive 700 hours of winter chilling or less should be considered extremely high risk. Zones with 700–850 chilling hours present moderate to high risk, and those with 850 or more chilling hours present low to moderate risk.

Extremely High Pressure Areas. If you are in a very high PD risk area, it is strongly recommended that tolerant or resistant varieties be planted. Before undertaking any new vineyard venture, talk to winery owners about their specific grape needs and make variety choices accordingly. If susceptible varieties are planted, be sure to isolate them from other blocks of tolerant varieties.

High to Moderate Risk Areas. Pierce's disease will definitely be a limiting factor to the production of susceptible grape varieties. Even with superior management, high levels of vine loss may be encountered. Site selection plays an extremely important role in mitigating this risk.

Moderate to Low Risk Areas. Moderate risk zones may still have relatively high disease pressure, so, again, site selection can help mitigate relative risk. In some northern areas where the presence of Pierce's disease has been confirmed, it may be more of a chronic problem rather than an acute one, but vine losses are still possible. Rouging strategies may be less aggressive in low risk zones as compared to areas of relatively high risk.

2. Select Site with Risk Management in Mind. Avoid planting vineyards near native perennial vegetation. Because xylem-feeding insects prefer a diversity of vegetation well supplied with water, creek and river bottoms pose an inherently higher risk. There is no set distance from a bottomland site that is considered safe. The farther away, the better.

3. Create Buffer Area. When selecting an area for a vineyard, be sure to have control of the vegetation several hundred feet in any direction from the prospective site. It is important to have the ability to manage vegetation adjacent to the vineyard. Removal of perennial trees and shrubs and mowing of fields will keep xylem-feeding insects from colonizing areas in proximity to susceptible vines. As with site selection, the greater the distance from perennial or unmowed areas, the better.

4. Remove Suspected Supplemental Vector Hosts. Wild grapevines can be colonized by *Xylella fastidiosa* but seldom show typical symptoms of Pierce's disease. Ideally, wild vines should be removed to the greatest distance that is practical. Become familiar with other plants capable of hosting *Xylella* and take steps to remove these plants.

5. Use Neonicotenoid Insecticides. After planting, apply imidacloprid or other nicotenoid through the drip system. First- and second-leaf vines can be treated with half the full-labeled rate. Treat third-leaf and older vines with the full-labeled rate. Become familiar with logistical practices and timing of effective insecticide application.

6. Learn to Identify Insect Vectors of Pierce's Disease and Monitor Vector Presence and Seasonality. More than thirty-five species of Pierce's disease vectors occur across Texas and the southeastern United States. Become familiar with what they look like, and use yellow sticky traps to monitor vectors in and around the vineyard. Trapping vectors can help determine when they are present, help identify where they are coming from, and lead to better management of vegetation types that host the numerous insect species that spread Pierce's disease.

7. Maintain Superior Vineyard Floor Management. Grapevines are not necessarily the favorite dining spot for xylem-feeding insects. Some

▲ *Cuerna costalis,* one of many insect vectors of Pierce's disease.

▶ Yellow sticky cards, which have a surface that traps insects, are placed on stakes in and around vineyards to help monitor PD vector activity.

insects, such as sharpshooters, need to change feeding hosts frequently to meet their dietary needs, and having a vineyard with weeds favors infestation by sharpshooters. The recommendation is to maintain a three- to four-foot weed-free area under the vines and to maintain vineyard row centers with close, frequent mowing.

8. Keep Vegetation Surrounding the Vineyard in Check. For the same reasons that superior vineyard floor management is important, vegetation around the vineyard should be mowed frequently to keep sharpshooter populations low. Allowing adjacent fields to grow for any length of time will attract a wide range of insect species. Infrequent mowing invites these populations to expand their range into the vineyard to find new food sources. Timely mowing beginning in late winter can prevent nearby populations of vectors from becoming established.

9. Become Familiar with the Symptoms of Pierce's Disease. Disease symptoms can change subtly from one grape variety to another. Become familiar with the visible symptoms of the disease, and be prepared to take action. Hoping for a cold, curative winter is not advisable.

Irregular leaf scorching is one of the first symptoms of Pierce's disease. To help diagnose the presence of disease, growers should become familiar with other symptoms as well.

10. Submit Grapevine Tissue from Suspected Infected Vines for Laboratory Analysis. Contact your local extension agent or extension specialist for help in identifying a laboratory that can conduct appropriate diagnostic tests for PD. Other pathogens or environmental stresses may produce symptoms similar to PD, so it is important to confirm vine infection status. Although it is economically impractical to submit every suspect vine for laboratory diagnosis, it is very important to use a diagnostic lab to

Vines suspected of having Pierce's disease should be tested by qualified diagnostic labs to confirm disease presence.

confirm suspected symptoms until you become confident of your ability to diagnose the disease in your vineyard.

11. Follow Vine Rouging Strategies Appropriate for Your Production Area. Relative risk of disease spread varies across growing regions. Learn which rouging protocol is appropriate for your area, and act immediately upon confirmation of vine infection. East of the Rocky Mountains, sharpshooter species can rapidly spread the disease within a vineyard. Removal of disease sources is essential to managing the epidemic.

A comprehensive Pierce's disease management guide for Texas grape growers, *Pierce's Disease Overview & Management Guide: A Resource for Grape Growers in Texas and Other Eastern U.S. Growing Regions,* is currently available on the Aggie Horticulture website, as well as various other sites. Most search engines should quickly provide a site for publication download. Grape growers in any region of Texas need to be thoroughly educated on site selection, cultural practices that reduce disease risk, and vector identification and management.

Cotton Root Rot

Cotton root rot is a fungal soil-borne pathogen endemic in much of Texas, southern New Mexico, and Arizona, as well as northern Mexico. Commonly called Texas root rot by the misinformed, *Phymatotrichopsis*

The first onset of cotton root rot. Vines typically die very quickly and can retain scorched foliage after death.

omnivora (previously *Phymatotrichum omnivorum*) has been widely recognized as a major limiting factor to grape growing since the late 1800s and has been the subject of numerous scientific studies since the 1920s. Many of these studies sought to identify resistant rootstocks, but results have been inconclusive and at times contradictory. Likewise, no fungicide tested to date has any proven ability to control this disease.

The pathogen has a host range exceeding twenty-five hundred cultivated plants and can remain viable in the soil for decades until a susceptible host is available. While grapevines are not the most susceptible perennial crop, the risk must be acknowledged where it occurs because it is capable of causing substantial losses. The pathogen typically favors alkaline soils, but isolated cases have been documented on neutral to slightly acidic soils. The accompanying map, originally produced by Stuart Lyda, shows the portion of the state thought to be at highest risk of vine loss.

The disease is not known to occur above the Caprock Escarpment and is limited to the east by acidic soils and to the west by soils with significant sodium content. Elsewhere within this affected region, areas with alluvial soils are thought to be at the highest risk of harboring the disease. Although the disease is known as cotton root rot, it is a common misconception that areas previously planted in cotton are at higher risk

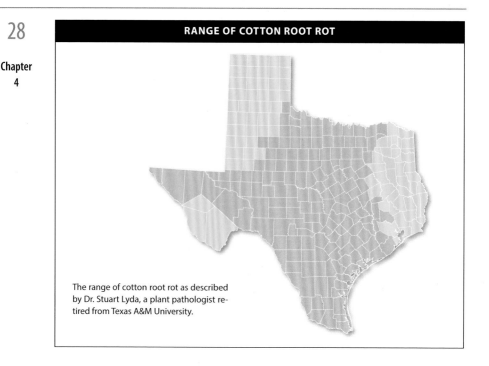

RANGE OF COTTON ROOT ROT

The range of cotton root rot as described by Dr. Stuart Lyda, a plant pathologist retired from Texas A&M University.

of harboring the pathogen. The name simply comes from the fact that the disease was first recognized and studied on cotton plants. It is postulated that many grapevines grow and produce fruit even though they are infected with the disease. Root systems with vigorous growth appear to be able to outgrow the pathogen in many instances. Disease pressure appears to be much higher under extremes of soil moisture content. Both drought-affected and waterlogged soils appear to exacerbate disease development.

Soil tests cannot determine the presence or absence of cotton root rot. When, in the 1980s, significant acreage in the Texas Hill Country was planted in apple orchards, cotton root rot devastated many of these plantings. Numerous attempts to control the disease by adjusting the soil pH or by introducing bio-control organisms failed to mitigate tree losses. Apples are among the most sensitive crops, and there is much to be learned from the attempts to control cotton root rot in apple orchards. Some orchard managers applied as much as five tons of elemental sulfur per acre in an attempt to lower the soil pH. The net result was that in the area where sulfur was incorporated the soil pH dropped to below 4.0, but the pH of the soil below the zone of sulfur additions remained above 8.0. The exchange capacity, or the ability of a soil to resist pH change due to high levels of alkaline calcium bicarbonate, makes changing the soil pH impractical. Because grapevines have very deep roots, pH change needs to take place throughout the root zone.

Beneficial fungi or bacteria can theoretically help control cotton root rot in shallow-rooted annual crops, but introducing those organisms deep into the soil profile in concentrations capable of competing with the cotton root rot organism is impractical. The ultimate control will be the identification and use of specific rootstocks or the identification and use of a systemic fungicide.

Self-rooted grapevines (grown on their own rootstocks) appear to be the most susceptible to infection and loss from cotton root rot. Rootstocks with *Vitis berlandieri* parentage, such as 1103P, 5BB, 5C, and similar varieties, appear to provide some protection, but under high disease pressure, even these stocks can fail. While early rootstock trials have yielded conflicting results, Dogridge and Champanel appear to be the rootstocks with the most consistent resistance to the disease. Both of these rootstocks have their own drawbacks, however. Dogridge probably resists cotton root rot by outgrowing it, but high vigor imparted to fruiting varieties can produce excessively vigorous vines with inferior fruit quality. Dogridge is also notorious for producing vines with excessive suckering. Combine these two characteristics with the fact that this rootstock variety is difficult to root, and the result is low availability from nurseries and less than stellar cultural characteristics. Champanel is a lower vigor vine, but it is very susceptible to root knot nematodes. Champanel is also tolerant of Pierce's disease, and no disease-free sources of plant material are currently available. Past attempts to graft PD-susceptible scion varieties onto Champanel rootstocks have resulted in as many as 90 percent of the vines showing PD symptoms in the first year. Champanel has been submitted to the Foundation Plant Services program at the University of California, Davis, and pathogen-free plant material should be available in the not too distant future. Careful selection of nursery sites in Texas will be required to keep vines grown in that region free from Pierce's disease

At the writing of this book, two trials are under way that may hold promise for cotton root rot control. Extensive rootstock trials have been planted in several locations with cotton root rot pressure in Texas. These new trials, which include new introductions from the program led by Andy Walker at UC Davis, should provide insight into potential resistance of new and existing rootstock selections. In addition, a new triazole fungicide, flutriafol, has recently been shown to have control capabilities

Wild vines, such as this *Vitis berlandieri* example, evolved with cotton root rot and are being used by some breeders to improve disease resistance in new commercial rootstocks.

in cotton, and a Section 18 label (the Environmental Protection Agency designation for temporary emergency approval) has been issued for that crop. Fungicide trials are also being initiated at two vineyards in the Texas Hill Country that have a history of cotton root rot problems. Between these two sets of experiments, it is hoped that some additional control measures will be developed and available to growers in the near future.

Weather

Advances in science can help overcome the threats from insect and disease outbreaks, but there is little hope that extreme weather events will be made manageable. When investigating "average" weather conditions, it is important to remember that those averages are made up of extremes. An area that on average receives twenty-eight inches of rainfall a year may in some years receive more than sixty inches and in other years under ten inches. Understanding the limitations of the individual areas of Texas can help you realize what challenges you will be facing, and choosing sites wisely can help you, in part, overcome these challenges.

Rainfall

While rainfall is generally helpful in supplying the entire vineyard floor with moisture and healthy root growth, excessive precipitation can present problems for the health and survival of grapevines, affect vine vigor, and seriously impact fruit ripening and grape quality. Canopy wetness drives most fungal pathogens, so tropical rain events can cause relentless disease pressure. In an ideal situation, a site that receives twenty-four to thirty inches of rainfall in a year, with an absence of significant rain from véraison, or the onset of ripening, to harvest, would produce healthy vines and high-quality fruit. In Texas, or any growing regions east of the Rockies, rainfall can vary widely from year to year. Grapes can be successfully cultivated across a wide swath of the state, but, when choosing a site, be aware of the weather conditions a region may experience and be prepared to deal with extremes that may occur over the years.

Spring Frost

Spring frost is perhaps the most heartbreaking weather event because it can wipe out an entire crop in a single night. As temperatures warm in the spring, buds begin to grow into shoots. Temperatures as high as 35 degrees Fahrenheit can result in frost formation that causes vine injury, and temperatures below 28 degrees Fahrenheit typically result in the loss of most if not all new green shoots and may even injure woody tissue.

Pruning techniques such as double pruning can reduce the risk growers face, but the most effective way to reduce spring frost risk is to choose a site that optimizes the flow of cold air away from vines. Planting on elevated sites that allow colder, heavier air to move to lower lying areas is the single most effective way of mitigating injury from spring frost. Some

Frost can cause severe injury to new shoots in the spring, resulting in vine damage and crop loss.

growers have opted to use overhead sprinklers as a means of reducing the impact of spring freezes. Spraying relatively large amounts of water to all parts of the vine prevents spring frost and freeze damage because the freezing of water is an exothermic reaction, or one that produces heat. Application of water must begin well before temperatures reach the freezing point, and, when used correctly, sprinklers can greatly reduce losses from spring frost. A thorough awareness and understanding of impending meteorological conditions and events is necessary to successfully mitigate frost damage. Incorrect water application rates or timing can under certain conditions actually make frost injury worse than might otherwise have been experienced.

Winter Freeze Injury

Winter injury differs from frost injury in that it can occur from late fall through spring and can kill or injure woody, permanent parts of the grapevine. The acclimation process and cultural practices to mitigate winter injury are explained in other portions of this text, but it is important to note that some parts of the state have an inherently higher risk of winter injury than others. Because of the moderating effects of large bodies of water, the Gulf Coast region rarely experiences weather events that cause winter injury. Conversely, the Texas Panhandle and many parts of

Overhead sprinklers have been used successfully by some growers to mitigate frost injury, but, if not used correctly, they can actually increase injury by cold temperatures.

Winter freeze injury can cause severe cordon and trunk injury, requiring the retraining of vines with newly emerging shoots.

West Texas have a significantly higher risk of experiencing winter conditions that may cause lethal vine injury. Specific varieties also vary in their ability to withstand low temperatures. Many varieties grown in colder parts of the world may be more cold hardy than others, but most are not adapted to ripening in the heat found in most parts of Texas. Other varieties, such as 'Cabernet Sauvignon,' are notorious for being "late acclimaters" in that they become fully hardy, or capable of withstanding the coldest winter temperatures, in midwinter and are more susceptible than other varieties to freeze events in late fall or early winter.

Hail and Wind

Spring or summer hail can cause serious injury to foliage and fruit. Although possible across the entire state, the High Plains, the Texas Hill Country, and most of Central and North Texas are at greater risk than more eastern or coastal growing locations. Both hail and high winds can do serious damage to vineyards at any time of the year. Straight line winds can shear off developing shoots and force the vine to regrow from secondary or tertiary buds, thus destroying the crop and having a severe

Hail damage can range from injured fruit to complete crop loss from spring or summer storms.

impact on vine vigor and health. Large hailstones can open wounds and provide entry access for bacterial and fungal wood-infecting pathogens, thus causing damage even during the dormant season. Efforts to reduce hail by using ground-based hail cannons that use sonic waves to break up hailstones in the clouds before they fall to earth have been largely ineffective, and the use of hail netting has not yet proven to be economically feasible.

CHAPTER 5

Systematics of the Genus *Vitis*

NY DISCUSSION of *Vitis* taxonomy or systematics must first acknowledge that botanists, not viticulturists, assign names to grapevine species. Among the former discipline, there exist two distinct groups of scientists, colloquially known as "lumpers" and "splitters." Lumpers look at subtle differences between groups of plant communities and find reason to lump them together as a single species, while splitters tend to view subtle differences as significant enough to require assignment of a new species name, thus assigning speciation more freely. The classical definition of speciation is predicated on the tendency of plants of a given description to preferentially reproduce among themselves in nature. This is difficult to prove without exhaustive study, so the nomenclature within this genus *Vitis* remains a speculative venture. Given the scientific tools available today, this science should become clearer in the not too distant future. The overview presented here is far from exhaustive but lists the most noteworthy species of economic importance or genetic potential for scion and rootstock breeding.

Within the family Vitaceae, it is commonly accepted that there exist fourteen genera and as many as nine hundred species. This family of plants can be found extensively throughout the tropics and subtropics, with ranges extending into northern pockets of temperate regions throughout the world. Plants classified within the genera *Parthenocisis, Ampleopsis,* and *Cissus* are common in nature and in ornamental horticulture, but the only genus with any food value is *Vitis.* The genus *Vitis* is divided into two subgenera: *Vitis,* species of which have thirty-eight chromosomes (2n = 38) and constitute the bunch grapes or "true grapes" that represent most of the naturally occurring species and widely distributed cultivars; and *Muscadinia,* species of which have forty chromosomes (2n = 40).

Muscadinia

While modern taxonomists now consider *Muscadinia* to be a sub-genus of *Vitis,* at one time *Muscadinia* was used as a distinct genus name. In addition to having a different chromosome number than other *Vitis,* the single member of this sub-species is morphologically different from other grape species. Muscadines have high resistance to soil-borne pests and foliar pathogens and have proven useful as sources of resistance in breeding programs looking to overcome these limitations.

Vitis rotundifolia

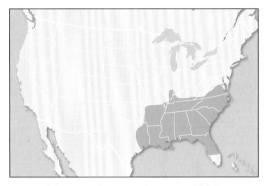

Range of the muscadine grape (*Vitis rotundifolia*).

The muscadine grape (*Vitis rotundifolia*) is an American species found in well-drained, moist, shady sites in a range that extends from eastern Texas, Oklahoma, Arkansas, and Louisiana, eastward to Florida, northward along the Atlantic coastal plains to central Virginia, and westward to Missouri and Kansas.

In addition to the differences in ploidy, or nucellar chromosomes, this subgenus differs from the *Vitis* anatomically. With muscadines, tendrils are not branched, bark does not exfoliate, and no diaphragm exists at the node, making the pith continuous. Muscadines are considered graft-incompatible with other grape species, and natural hybridization occurs only with *V. munsoniana.* Artificial hybridization has been accomplished with other *Vitis* species. Like other wild grape species, *V. rotundifolia* is dioecious, so both male and female plants are needed in close proximity for fertilization to occur. Like cultivated *vinifera* cultivars, modern muscadine varieties are hermaphroditic and do not require pollinators.

Muscadine leaves are all alternate and either oval or broadly cordate, with margins that are seldom lobed. Leaves are leathery, with a glabrous, dark-green upper surface. Muscadines bear fruit from July to September, with the grapes dropping singly from the rachis when ripe. Fruit may be sold fresh, or pressed and sold as juice, or fermented into wine. Because of their superior insect and disease resistance, they have been used as a source of genes for improvement of *vinifera* cultivars and hybrid rootstocks. Muscadines serve as a genetic source of tolerance to root knot and dagger nematodes, phylloxera, and Pierce's disease. In addition to *rotundifolia* muscadine grapes, two other species are commonly grouped in the *Muscadinia* subgenus: *V. munsoniana,* found in subtropical Florida and parts of the Caribbean, and *V. popenoii,* found in Mexico.

Vitis

There are perhaps fifty to sixty species within the *Vitis* subgenus, and the native range of these species extends from North America and the Caribbean eastward to the Middle East, and through China and Japan. Taxonomists disagree on the number of species within North America, with some arguing for as few as eighteen and others, as many as thirty.

Vitis aestivalis

The summer grape (*Vitis aestivalis*) grows in dry woods and thickets and along roadsides in a range that extends from Oklahoma, Arkansas, Texas, and Louisiana eastward to Florida, northward to New Hampshire, and westward to Wisconsin and Kansas. Some taxonomists group it together with what others name *V. lincecumii.* One economically important cultivar, 'Norton' ('Cynthiana'), is thought to be wholly derived from *V. aestivalis,* and some believe that the lineage of 'Black Spanish' is a complex hybrid that includes *V. aestivalis.*

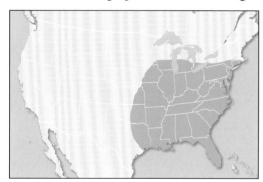
Range of the summer grape (*Vitis aestivalis*).

V. aestivalis is a strong grower, with leaves that somewhat resemble those of *V. labrusca.* Leaves of *V. aestivalis* are simple, alternate, and generally cordate-ovate in shape. Fruit is borne from September through October and is dark blue to black with a thin bloom. Although valued as a commercial fruiting cultivar, *V. aestivalis* lacks phylloxera resistance but is used in hybridization as a source of powdery mildew resistance.

Vitis arizonica

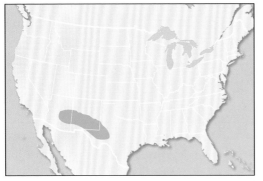

Range of the canyon grape (*Vitis arizonica*).

Commonly known as the canyon grape, *Vitis arizonica* can be found in ravines and gulches at altitudes between 2,000 and 7,500 feet in parts of western Texas, New Mexico, and through central Arizona. Two subspecies of this species have been described. *V. arizonica* var. *glabra* has shiny, smooth, nearly glabrous leaves, while *V. arizonica* var. *galvinii* has larger, more serrate or lobed leaves and larger fruit.

This vine is weak in growth and has a slender trunk tapering rapidly in diameter from the base to the apex of the plant. Fruit matures from July through August in clusters usually shorter than the leaves. This species may hold tremendous promise as a source of tolerance to Pierce's disease because the genetic basis for Pierce's disease resistance in this species is due to a single small cluster of genes. Forms of this species also have excellent resistance to dagger nematodes (*Xiphinema index*).

Vitis berlandieri

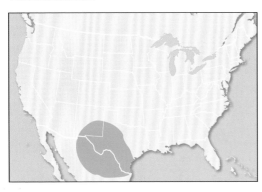

Range of the winter grape (*Vitis berlandieri*).

Commonly known as the winter grape, *Vitis berlandieri* is generally found on calcareous soils but sometimes on well-drained sandy sites. This species is abundant on the limestone hills and creek bottoms of Central and southwestern Texas, from west of the Brazos River to the Rio Grande and into Mexico from Coahuila to Veracruz. The vine is stocky, branched, and can be an aggressive climber. Leaves are generally cordate-ovate, often broader than long, with a sinus ranging from narrow to broad. Fruit ripens between August and October, with the rachis as long as six inches.

This species was historically important in the development of the first phylloxera-resistant rootstocks after the introduction of this New World pest into western Europe caused the collapse of the French wine industry. *V. berlandieri* is also a genetic source of tolerance to powdery mildew, downy mildew, drought, salinity, and iron chlorosis. It is known to freely hybridize with other species in its native range. Graft failure is common on wild vines, and it is difficult to root.

Vitis cinerea

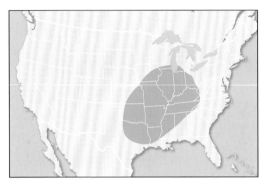
Range of the sweet winter grape (*Vitis cinerea*).

The sweet winter grape (*Vitis cinerea*) is a high-climbing species that is found in moist alluvial soil along streams, thickets, and bottomlands from Texas, Oklahoma, and Arkansas eastward to Florida and northward to Missouri, Indiana, Illinois, and Wisconsin. Leaves are from four to eight inches long and are subrotund to cordate in shape. Fruit, usually a dull black in color, ripens from August to October and has a pungent flavor.

V. cinerea is poorly adapted to high-pH soils but has good resistance to powdery mildew, downy mildew, and black rot. The resistance of this species to these fungal pathogens would be of greater value in cultivar breeding if the fruit were of higher quality. It may be especially important to rootstock breeders because it has tolerance to *Xiphinema* nematodes and phylloxera, but it is difficult to root and graft failure is common.

Vitis labrusca

The fox grape (*Vitis labrusca*) is a strong-growing vine, highly adapted to acidic soils, and found in sandy, alluvial soils from Arkansas east to Georgia and north to Massachusetts, Ohio, and Indiana. *V. labrusca* leaves are alternate, each one opposite a forked tendril or a flower cluster, and they are thick and ovate to cordate-ovate in shape. The upper leaf surface has a dull green appearance, and the lower surface is covered

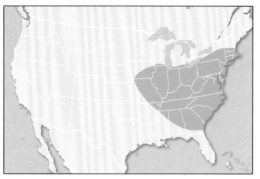

Range of the fox grape (*Vitis labrusca*).

in dense white to reddish-brown pubescence. Fruit ripens from August to October, and clusters are often as broad as they are long, usually with no more than twenty berries per cluster.

Numerous economically important cultivars, such as 'Concord,' are derived from *V. labrusca* × *V. vinifera* hybrids and are used for table fruit, juice, and wine. *V. labrusca* is a genetic source of cold hardiness and large berries but has a distinct flavor that some find objectionable. This species has some resistance to powdery mildew, anthracnose, and flavescence dorée. Susceptibility to phylloxera (probably due to *V. vinifera* parentage) and intolerance of calcareous soils are among its weaknesses.

Vitis mustangensis

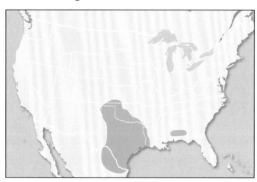

Range of the mustang grape (*Vitis mustangensis*).

Previously named *Vitis candicans*, *V. mustangensis*, or the mustang grape, is a very vigorous, high-climbing species, with old vines at times attaining a length in excess of forty feet and trunks a foot or more in diameter. The mustang grape is common in North, East, and Central Texas and ranges to southwestern Texas and northern Mexico, eastward into Louisiana, and north to Oklahoma and western Arkansas. It is found along rivers, creeks, and roadways. An adjunct population of *V. mustangensis* can also be found in southern Alabama, although some taxonomists believe it to be *V. shuttleworthii*.

A striking characteristic of this species is the inverted saucer shape of the leaf blade, which is borne on the petiole in such a manner as to provide a shingled canopy effect when the foliage is dense. Leaves are two and one-half to five inches long and wide and cordate-ovate to reniform in shape, while the lower surface is covered in a felty white pubescence. The fruit matures from late June to August and may persist until late autumn. The black fruit are borne in clusters of from three to twelve berries that can each approach an inch in diameter.

V. mustangensis has good resistance to phylloxera and is well adapted to hot, dry conditions. It also has very good resistance to black rot and powdery mildew, root knot nematodes, and Pierce's disease. It is not used as a rootstock because it is difficult to root, is frequently graft-incompatible, and shows poor tolerance of high-lime soils. *V. mustangensis* also freely hybridizes in nature.

Vitis riparia

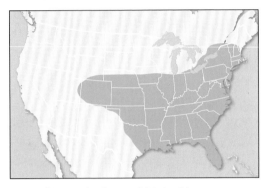
Range of the riverbank grape (*Vitis riparia*).

Vitis riparia, or the riverbank grape, is a moderately vigorous grower attaining considerable height, especially in rich bottomlands. The name *riparia* means "from the river," which alludes to its most common habitat. It is the most widely distributed of North American grape species, ranging from Florida to New England and west to the Rocky Mountains. Leaves are opposite, with blades ranging in size from two to six inches in length and up to four inches in width. The leaves are cordate-ovate in shape, approaching triangular at the apex, with long sharp teeth and shoot tips that are wrapped in the young leaves. Fruit ripens from August to October and may persist on the rachis into late fall. Berries are dull black in color and commonly pea-sized, with fruity but acidic flavor.

V. riparia is a common parent in rootstock breeding because of high phylloxera resistance and winter hardiness. It is a shallow-rooted vine, and rootstocks derived from this species are consequently quite devigorating. It roots well and is easy to propagate but is not tolerant of high-lime soils. For scion breeders, it offers a source of resistance for black rot and powdery and downy mildews, but some forms are subject to very early budbreak in the spring.

Vitis rupestris

Range of the sand grape (*Vitis rupestris*).

Commonly known as the sand grape, *Vitis rupestris* is quite diminutive in stature, resembling a small shrub, and rarely exceeds three feet in height. It most commonly has a prostrate growth habit and rarely climbs. While the range of this species once extended from southwestern Texas to Pennsylvania, it may now be limited to southern Missouri.

V. rupestris once grew on sand hills or calcareous soils along streams or gulches, often at the head of ravines, but it is now found primarily on coarse-textured to gravelly soils. Leaves average three to four inches in width and are usually broader than long. Leaves are reniform to reniform-ovate in shape and fold upward to expose the pale green lower surface, and petioles are a distinctive red or purple. Fruit is borne from June to August, dropping at maturity. The black berries are approximately one-quarter inch in diameter and have a very thin skin, with deep purplish or black juice.

V. rupestris is known for its high vigor, drought tolerance, and strong resistance to phylloxera. The vigor is a result of a deep root system that holds the vine in place in the rocky creek beds it inhabits. Because of the prostrate growth habit and small stature of *V. rupestris,* it is extremely susceptible to browsing by animals and thus rare in nature. It roots easily and is a common parent of commercially important rootstocks, and the rootstock St. George is wholly derived from this species. Its weakness as a stock is its intolerance of high-calcium soils. Scion breeders use this species to gain resistance to black rot and flavescence dorée. There are conflicting reports on downy mildew resistance in *V. rupestris,* which probably signifies variability within the species.

Vitis vinifera

By several orders of magnitude, the *Vitis vinifera* species has the greatest economic value and is the most widely cultivated around the world. The name *V. vinifera* is commonly used in current nomenclature, but

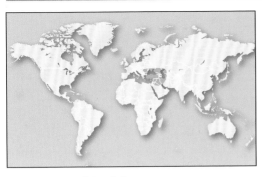

Ancestral range of *Vitis vinifera*.

43

Systematics
of the Genus
Vitis

the ancient wild types are divided into two subspecies: *V. vinifera sylvestris*, of southern and central Europe, northwestern Africa, western Turkey, and Israel; and *V. vinifera caucasia*, found in Ukraine, Moldova, southern Russia, Armenia, Turkey, Iran, Central Asia, and the northwestern sections of the Indian subcontinent. In the latter subspecies, most fruit are small and seedy, but larger fruit forms are found in feral vines. Leaves are quite variable in size and shape but are generally glabrous with waxy cuticles. Fruit ripens over a lengthy period of time and features tremendous variability in size, color, and flavor characteristics.

First and foremost, the genetic variability of this species is useful for cultivar improvement. Most of the common wine varieties are of ancient origin. Hybridization with American species has primarily been explored in an attempt to incorporate cold hardiness or resistance to fungal pathogens. *V. vinifera* is used in rootstock breeding to increase drought tolerance and adaptation to high-pH soils, but the limiting factor of stocks with *V. vinifera* parentage is susceptibility to phylloxera.

Other Species

There are many more grape species from various parts of the world that may ultimately assist breeders in overcoming limitations of existing cultivars and rootstocks. Included in this group are *V. amurensis*, native to southern China, Japan, and areas southward to Java, *V. caribaea*, native to Central America, Venezuela, and the Leeward Islands of the Caribbean, and *V. monticola* from the Texas Hill Country. The preservation of this germplasm is a crucial resource for the next generation of plant breeders seeking to solve the problems that the future will ultimately bring.

Choosing a Vineyard Site

T IS probably safe to say that among extension horticulturists who work with perennial fruit crops there have been more than a few who rarely left their offices to investigate a problem a grower was having in the field. Their reasoning might be that 90 percent of the problems growers face are caused by the site or, more specifically, the poor selection of a site. With that assumption, one might then argue that once the vineyard or orchard has been planted, there is nothing anyone can do to overcome the limitations imposed by that decision. Most vineyard advisors do not really feel that way, but the point needs to be made. A decision to plant a particular site is a long-term commitment that must be made for all the right reasons. There are a number of factors to consider in site selection; understand them all and choose wisely to minimize risk and maximize your potential for success.

The Role of Soil Sampling

One of the first steps a new grower with a prospective site should take is to get a soil sample tested. It is not a bad idea, but there are a number of things a soil sample test will tell you and many more that it will not.

Soil sample tests will tell you the relative acidity or alkalinity of a soil and give you an analytical measurement of the relative nutrient content of that soil. Of those test results, soil pH is perhaps the most widely reflective of the challenges a grower will face. The pH scale runs from 1 to 14 and is a logarithmic reflection of the relative concentration of hydrogen ions (H^+) or hydroxyl ions (OH^-) in the soil solution. However, the range of soil pH levels capable of supporting healthy plant life is functionally between 4 and 9. At a pH of 7, the solution is considered neutral; there are equal amounts of H^+ and OH^- in solution. A numerical rise to 8 indicates the presence of ten times more hydroxyl

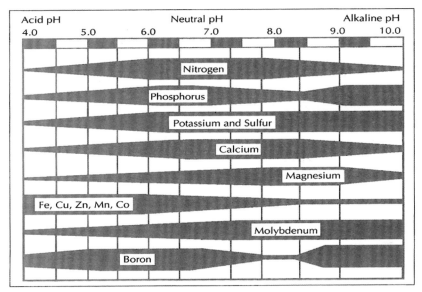

Soil pH levels can affect nutrient availability to plants.

than hydrogen ions in the soil, and, conversely, a soil with a pH of 6 has ten times more hydrogen than hydroxyl ions. Lower pH soils are considered acidic, and higher pH soils are termed alkaline.

The primary importance of soil pH is the impact it has on nutrient availability to plants. An element may be present in a soil, but a high or low soil pH may inhibit the ability of some plants to absorb it. Nutrient availability charts are useful resources for showing average plant response, but it is important to remember that actual performance will vary across a range. *Vitis labrusca* × *V. vinifera* varieties such as 'Concord' and 'Niagara' are native and adapted to the very acidic soils of New York and New England and perform quite well at a pH of 4.5. Conversely, what we now deem as descendants of *Vitis vinifera* evolved across a vast area that includes southern and central Europe, southern Russia, Armenia, Israel, Turkey, Iran, Central Asia, and the northwestern sections of the Indian subcontinent, and these plants prefer soil types like those of their place of origin: near neutral to slightly alkaline soils. In acidic soils, *vinifera* vines suffer from a condition known by the German word *sauerschaden,* which is actually a toxicity due to increased aluminum availability in acidic soils. In high-pH soils many grapevines, especially those of eastern US origin, suffer from a lack of iron and zinc availability. Well-chosen rootstocks and, to a measured extent, chelated soil amendments can overcome these physiological characteristics, but it is important to understand where grapevine species are well adapted and where they are not.

Routine soil testing cannot tell you how deep soil profiles are, cannot adequately describe how well soils are drained, cannot tell you if there are physical factors that may limit grapevine growth, and cannot tell you if there are or are not soil-borne pathogens that may kill your vines. Routine sampling will also not tell you if other farming enterprises have left residual chemicals that may impede vine growth. Growers can, however, conduct specialized soil tests to determine the presence and concentration of harmful nematode species.

Soil samples are a snapshot of a geological medium in which you will attempt to grow a perennial crop. If a grower wants to track vine/site/seasonal interactions, petiole sampling will give a grower a much more dynamic understanding of how to respond to the seasonal fertilization needs of a vineyard. A thorough understanding of the threat of soil-borne pathogens is essential to proper vineyard planning. Consult other chapters in this work for more detailed information on both the pathological and the nutritional limitations that will affect vine growth. Guidelines for taking soil samples can be found in the chapter on grapevine nutrition and vineyard fertilization.

Soils in a given field can be quite variable, and separate soil samples of different soil types should be taken independently.

Soil Structure and Ion Exchange

Soils need to be viewed as a dynamic entity with both physical and biological properties. Soils consist of combinations of three types of soil particles (sand, silt, and clay), in different sizes and shapes. In relative terms, sand particles are approximately the size and shape of a beach ball, a silt particle is comparable to a Frisbee disk, and a clay particle approximates the size and shape of a dime. Soils with high amounts of sand are considered coarse and have relatively large air spaces and relatively small amounts of soil particle surface area. Conversely, soils with higher concentrations of silt and clay are considered heavier and generally have more surface area and less air space.

The surface area of each soil particle has exchange sites that are negatively charged. Positively charged soil nutrients are attracted and held to these exchange sites, while negatively charged ions are not and thus remain free in soil solution. Silt and clay particles are much smaller than sand particles, have greater surface area, and consequently have more sites to retain positively charged nutrients. Sandy soils are generally characterized as being well drained, with air returning to the soil quickly after a rain, but with less nutrient holding capacity than silty or clay soils. Heavier soils can have physical properties that mimic the larger air space when the clay soils are considered well structured. Again, think of clay particles as dimes. If they are piled on a table, all flat, then there is little air space between the coins in the stack. If the dimes could be arranged with some of them flat, others on edge, and more stacked on edge at various angles, one can imagine the greater air space that group of coins might have. Clay particles can congregate together in groups called colloids. As a group, soils with strong colloidal structure behave physically like coarse soils in that they have good drainage and sufficient air space for good root growth and function.

These well-structured soils with strong colloidal structure also have a higher number of exchange sites and are thus capable of holding a greater amount of positively charged nutrients. The relative ability of a soil to retain nutrients from organic matter is expressed in relative terms as the cation exchange capacity (CEC) of a soil. Organic matter is extremely important in soils in that it helps maintain good soil structure, holds soil moisture, facilitates soil drainage, and increases soil CEC. Most soils in Texas, as is the case across the southern United States, are very low in organic matter, so the use of cover crops and mulches can be beneficial in improving or maintaining properly structured soils.

An integral part of soil organic matter is the vast quantity and variability of microorganisms. Maintenance of this diversity, and the complex competition and balance between these organisms, is believed to play a significant role in reducing the incidence and severity of soil-borne pathogens. Many soil microorganisms are also important in the decomposition of organic matter and in the conversion of nutrients to a usable form.

Internal Soil Drainage

Grapevines require well-drained soils to remain healthy and productive. Water and air must be present in the root zone so that the roots can absorb water and nutrients. When soils become saturated with water, however, oxygen is driven from the air spaces and root function will cease until air is once again returned to the soil. Planting vineyards on a slope will help during short-term high rainfall events and enhance surface drainage of excessive moisture. On other occasions when rainy weather sets in for days on end, internal soil drainage will be necessary for vines to survive.

A simple test may be conducted to determine if internal soil drainage is adequate. To check for drainage, use a hand-held post hole digger and go down approximately thirty-six inches deep. Fill the hole with water. If the water in the hole drains within twelve hours, the site has excellent drainage. If it takes twenty-four hours to drain the hole, the internal soil drainage is adequate. If it is not drained within that time frame, it is probably best to avoid that site. Visual inspection of a soil profile can also be helpful. Typically, topsoil will be made up of sand, silt, and fairly well-structured clays. Subsoil horizons may be problematic, however. Subsoils that are red or brown indicate the presence of an aerobic environment (i.e., oxygen is present). Gray or yellow clay subsoils are indicative of a reducing environment (i.e., oxygen is generally not present), and root growth will not be supported.

Air Drainage

As with all perennial crops, the risk of crop loss from spring frost remains among the leading threats to consistent production from year to year. Unlike California and the Mediterranean region, most of the southern and central United States has a continental climate, which means it is subject to wide swings in both rainfall and temperature.

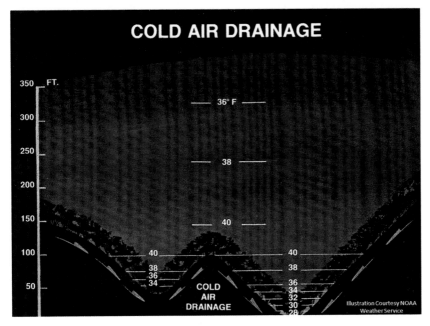

COLD AIR DRAINAGE

Cold air drainage helps protect vineyards during cold spells. Because cold air flows to lower levels, vineyards on higher elevations may be less subject to frost and freeze injury. (Courtesy National Oceanic and Atmospheric Administration Weather Service)

After vines have received adequate winter chilling, rest is broken with the advent of moderate to warm temperatures in the spring and growth begins. In the Texas climate, the lack of modifying influences commonly results in late spring cold fronts ushering in colder than anticipated frost events that kill tender shoots and clusters in grapevines. Peach growers expect a complete freeze-out will occur in one out of seven years, and for grape growers the odds are similar. No matter how ideal the soil conditions appear, creek bottoms and river bottoms should be avoided because they tend to be "frost pockets." A vineyard site in an elevated setting is far less prone to losses from spring frost than sites in lowland locations. On a frosty morning, cold air, which is denser than warm air, settles in low-lying sites, making them more susceptible to frost than sites in upland locations. In addition to mitigating frost risk, air movement in the vineyard facilitates the drying of the vineyard canopy, which can reduce the risk of fungal disease infection.

Herein lies the dilemma of the Texas Hill Country: where there are appropriate soils, there is no air drainage, and where there is air drainage, there are no appropriate soils. In reality, this conundrum means that the proper vineyard site in the Hill Country, or any other part of the state, is not the average piece of property one happens to own; it is the exceptional site the savvy grower chooses after long and deliberate searching.

For the High Plains, small increases in elevation can mean several degrees difference in temperature on a cold, still morning. In many other parts of the state, prospective growers can find elevated sites with excellent soils. Again, air drainage should be considered every bit as important as soil depth and drainage in selecting a site for a commercial vineyard.

Water Resources

Vineyards in every part of the state will require supplemental irrigation. Adequate water availability and quality needs to be verified before a site is planted. Municipal water may be capable of establishing a vineyard in its initial years, but as vines mature and water requirements increase, this source of water may well become cost prohibitive. Well drillers near the prospective vineyard site can frequently give you an idea of typical well pumping capacities and water quality in a given area. In some circumstances, prospective growers wishing to purchase a parcel of land to establish a vineyard have included a provision in the purchase contract that the sale of the parcel is contingent on the drilling of a well to deliver a specific quantity of water with acceptable chemical quality.

The general rule of thumb is that a water well needs to be able to provide five gallons per minute (gpm) per acre of vineyard. While others may argue that the capacity should be higher, a well pumping twenty-four hours a day could deliver in excess of eighty gallons per vine per day under typical spacing conditions (605 vines per acre). Given general recommendations that vineyards receive one acre-inch per week, which would require the application of forty-four gallons per vine per week, a five gpm/acre well capacity is adequate to meet those needs.

Water quality is a concern in all parts of the state. The greatest chemical limitation to potential vineyard irrigation water is high salinity. There are numerous measures of salinity, but electrical conductivity (EC) and sodium absorption ratio (SAR) are common expressions of salinity. As a rule, grapevines are more tolerant than many fruit crops, but water with greater than 1,000 ppm dissolved salts, an EC above 4 millimhos per centimeter (mmhos/cm), or an SAR above 7 will cause a serious reduction of plant growth and yield. Salts are water soluble and can be flushed out of soil profiles with high rainfall, but under drought conditions salts can build up in soils and limit the growth and productivity of a vineyard. While irrigation practices can be modified to address salinity issues, it is best to avoid sites with inherent problems.

Proximity to Wineries

Even if one finds the perfect spot to grow grapes, new problems arise if that site is five hundred miles from the winery that you want to process the harvest. Grapes are a nonclimacteric fruit, which means that they do not continue to ripen after harvest. From the point of harvest, grapes begin a decent into decay impeded only by temperature and time.

To produce premium quality wines, harvested fruit needs to be crushed and cooled as quickly as possible to control the parameters of fermentation. In practical terms, the quicker a crop is delivered to a winery, the better the potential outcome for a good wine. Quick processing is not an automatic guarantee of great wine, but to optimize wine quality there must be clear communication with the winery, dependable transportation must be arranged, and the winery must be prepared to act. There have been occasions when harvested fruit was left in half-ton capacity bins under "shade trees" in the middle of August for thirty-six or more hours, awaiting transportation. Fruit that was reportedly harvested at twenty-four degrees Brix (fermentable soluble solids in the fruit) was actually received in the winery at nineteen degrees Brix, two days after harvest was completed. Thus, between harvest and delivery, fermentation had already begun. Fermentation under uncontrolled, summertime temperatures is far from ideal, and the winery should not have blamed the grower for the diminished fruit quality when its transport system had not operated in a timely manner.

Understand what is happening to your fruit before *and* after harvest, talk with other growers, and deal only with wineries committed to producing the highest quality wine. Long-term successful grower/winery relations are built on integrity and fairness to all concerned. With all of these factors in mind, a grower will find that the closer a vineyard is to the crush pad, the lower the chances that fruit will be handled improperly.

Where You Live

As previously stated, site selection needs to take precedence in where a vineyard is located, but ready access to the vineyard is also important

Sites with appropriate soil depth and drainage, an ample supply of high-quality irrigation water, and superior air drainage to mitigate the risk of spring frost are very hard to find. Site selection is the most important decision a new grower will make.

in responding to immediate management needs. Good managers need to be able to turn on a dime in responding to rapidly changing vineyard and climatic conditions. Quick response time is impossible if one lives any appreciable distance from the vineyard. Many new growers may need to travel quite some distance to their vineyard site, and some will soon realize that travel time will be a limiting factor in their ability to accomplish tasks in the vineyard promptly. As the saying goes, the best fertilizer in the vineyard is a grower's shadow.

CHAPTER 7

Time line for Establishing and Planting the Vineyard

STABLISHING a successful vineyard involves progressing through seven critical steps, each of which requires sufficient investment of time and resources before proceeding to the next.

Planning

For some, researching all of the facets of successfully growing grapes—from determining overall goals, to site selection, variety and rootstock selection, and so forth—may take months if not years. As mentioned earlier, choosing a site is one of the most critical decisions a prospective grower can make. Deciding on vineyard size, envisioning potential for expansion, fully assessing risks and how to deal with them, and marketing all take time. As a rule of thumb, allow a fifteen-month lead time before anticipated planting in order to accomplish all of the pre-planting tasks. Rushing to meet an unwise early planting window usually results in problems that can plague you for the life of the vineyard operation.

Once the decision has been made to move forward with the time and financial commitment necessary to establish a vineyard on a particular parcel of ground, two time-sensitive tasks should be immediately addressed.

Procuring Plant Material

The first, and one of the most critical tasks, is the procurement of plant material. Care should be taken in choosing a reputable nursery that can supply true-to-type, disease-free grapevines in the variety and rootstock of your choice. This purchase may seem like an easy task, but a nursery should be contacted at least fifteen months before the anticipated planting date. Avoid the temptation of planting what the nursery has on hand rather than what is optimal for your vineyard. While nurseries do some speculative propagation and planting of some popular combinations, the only way to ensure that you get exactly what you want is to custom order grapevines.

Choosing and paying for certified plant material is highly recommended. There have been many cases in which growers take the time and effort to plant a vineyard but did not invest in certified disease-free grapevines, only to find out three years later that the majority of the vineyard was infected with leaf roll or another virus. While in the past the disease-free certification had not guaranteed freedom from bacterial pathogens such as the one causing crown gall, efforts are being made across the country to include bacterial pathogens in the certification program.

Site Preparation

The second critical task is site preparation, which should begin the spring or summer before the season you intend to plant. It is really best to choose a site that has been cleared of trees or other woody perennial shrubs, but, if some remain, they should be removed and the immedi-

Eliminating existing perennial vegetation is an essential part of site preparation for a new vineyard.

ate area root plowed to expose remaining woody root tissue. Every effort should be made to remove old roots from the entire site. Decay organisms that normally would not be problematic on new grapevines may reduce growth if they are present in abnormally high numbers because of other woody tissue that has been left at the site and is decomposing.

It is also common for new growers to encounter perennial vegetation such as Johnsongrass, Bermudagrass (native or coastal), or perennial ragweed on a site they intend to develop and plant. It is much easier to rid an area of serious competitive vegetation before a vineyard has been planted rather than afterward.

By starting this process in the year prior to planting, growers do not need to concern themselves with the unintended consequences of vine injury due to incidental herbicide contact. An approach that has worked is to apply a nonselective herbicide such as glyphosate across the entire area that one intends to plant. Application rates for these products are typically adjusted based on what target weeds are present. Applications of glyphosate in early summer may provide only limited suppression of perennial weeds but should control most if not all germinating broadleaf and grassy weeds. It is common practice to run a disk or other type of cultivator to incorporate the organic matter left behind by the glyphosate application. At this time, minor leveling of ruts or berms can be corrected to level the new vineyard floor. With additional rains, perennial weeds and an additional flush of annual weeds will normally occur. Successive glyphosate applications can continue the elimination of weeds, and late-summer and autumn applications typically yield the best results when targeting hard-to-control weeds.

Because poison ivy, Virginia creeper, Johnsongrass, and Bermuda-grass are especially difficult to eliminate, they should be allowed to produce mature leaves prior to glyphosate application. Repeat applications of glyphosate may be needed at one-month intervals throughout the fall in order to optimize the control of difficult vegetation. At this point, it is time to begin installing infrastructure.

Irrigation

If a water well is already in place, irrigation main lines, lateral lines, and risers should usually be installed the fall before planting. Above-ground polyethylene piping that delivers water through emitters is usually deployed immediately after planting.

Typically, supplies such as PVC pipe, fittings, and trenching tools are

Irrigation lines are typically buried twenty-four to thirty inches underground from the well to the risers at each row.

sourced locally. Transportation costs often eat up any savings that might be achieved by purchasing from distant sources. It is wise to shop for specific irrigation supplies such as timers, valves, emitters, injectors, and filters by researching national or regional companies that specialize in commercial agricultural irrigation.

If you have not had significant experience in construction projects that utilized water-pressurized piping, seek out local contractors who have hands-on experience with these projects. Once an irrigation pipe is buried, it can be frustrating and expensive to diagnose, locate, and repair leaks that might have been avoided by getting experienced help with installation.

Trellis Construction

Sourcing materials for a vineyard trellis takes an approach similar to that for an irrigation system. Items such as posts and wire may be most economically purchased locally because of the high costs of transportation. That is not always the case, however. Tractor trailer loads of steel t-posts and/or wire may be available from distant suppliers at a reduced rate, but utilizing this option may necessitate finding another grower who needs similar supplies and working out a deal to share the costs.

If t-braces or specialized cross-arms are being utilized in the trellis, finding a vendor specializing in those components will probably be a necessity. It is better to use materials specifically designed to withstand the wear and tear a trellis will encounter than to engineer components yourself.

Building a trellis is not rocket science, but it helps if you have significant fence-building experience to properly construct a durable, functional framework for your vineyard. Having straight rows is more than a matter of aesthetics; it will be crucial when moving equipment through a mature vineyard. A lopsided, irregular trellis can result in continual maintenance problems and may make the conversion to mechanical harvesting nearly impossible. Knowledge of geometric principles is essential in making sure that rows are exactly the same distance apart and that the whole block remains square. An error of an inch in one row will multiply itself as trellis construction progresses across a field, reaching the point where serious errors can be encountered at the end of the block. Utilize sound geometric principles in creating a technique for establishing

Having end posts and line posts in place prior to planting will make it much easier to complete the trellis installation without damaging young vines in the process.

vineyard block corners and laying out post positions and end posts. Trust your measuring devices, not your eye. Even though posts are in the proper locations, they may not be driven in exactly perpendicular to the ground, and relying on your eye to "square up" the block as you go will give you problems down the road. Again, having an experienced hand may prove invaluable even if you want to do the majority of the labor yourself.

It is traditional that at least end posts and line posts are in place prior to planting the vineyard. Once wires have been strung, moving equipment across rows becomes impossible and crews will then need equipment capable of digging holes at an angle underneath the wires. If vines are planted before any wires are installed, it is imperative that wires be stretched before growth begins in the spring. Rolling wires down the row and then stretching and attaching them to posts can result in unintentional removal of tender green shoots on young vines.

Planting

Many California nurseries will try to deliver plant material to you in May or even June. Planting of dormant vines at this time of year com-

Sturdy end posts and anchors are essential elements of a well-designed trellis.

monly results in poor growth if not outright disaster. Insist that dormant plant material arrive no later than the end of March, so that planting can proceed while temperatures are still relatively cool. Planting in February to March is ideal because, once planted and watered in, roots can begin to grow before buds force from the aboveground portion of the vine. This "jump" in root growth will improve a vine's ability to absorb the water necessary for vegetation growth and expansion of green tissue. In May or June, temperatures can easily climb into the upper nineties, which will greatly increase the transpirational need in a vine, a need that will not be met by a slow-growing root system.

Some nurseries will sell and deliver "green grafts," which are actively growing grafts that arrive in small rooted pots or plugs. Extreme care must be used in handling this type of plant material, but they are a viable option in vineyard establishment. These vines often dry out quickly because of small pot size and may need to be watered twice a day while awaiting transplantation. If green grafts arrive during hot weather, they may need to be acclimated in partial shade for three or four days to help them survive the extreme changes in temperatures. Partial shade, frequent watering of potted plants, and thorough watering after transplant-ing will be necessary for establishment of green grafted vines.

When nursery stock arrives, open the boxes and check to ensure that the plant material is what was ordered, that the count is correct, and that plant material has arrived in satisfactory condition. Pay special attention to moisture levels within boxes. Once bare-rooted plant material dries out, it is irreparably damaged and will not grow properly. If there is a problem, contact the nursery immediately and arrange for a settlement (they most likely will not have new plant material on hand to send you). After opening and inspecting the boxes, reseal and store them in a cool, damp place. *It is imperative that the roots of dormant vines stay moist but not saturated until they are planted in the ground.* A walk-in cooler with temperatures between 38 and 45 degrees Fahrenheit is ideal, but make sure the tempera-ture does not drop below that level so that nursery stock does not freeze.

It is common for new growers to want to share the excitement and the experience of the new vineyard by having their friends assist in the plant-ing activity. While friends may have the best intentions when trying to help, it may be wiser to hire a crew you are not afraid to correct if they are not doing the job to your satisfaction. There is a rule shared among grape growers that the number of your friends is inversely proportional to the number of planting parties you have. It is important to plant to the proper depth, keeping the graft union at least four inches above ground level even after soil settles around the new vine. Covering the graft union will result in the rooting of the scion (fruiting portion of the vine) and the rootstock effect will be lost. Vines should be planted as close as possible to the exact line created by the end posts and line posts. Over time, as vines age and grow in diameter, misplaced vines will create problems for the movement of equipment through the vineyard. Keeping standard spacing between vines in the row helps create a uniform and productive canopy.

Before planting day, holes for vines should have been pre-drilled; if not, a digging crew should precede the planting crew. Locations where the soil has a high clay content are best planted with holes dug the same day, to keep clod formation to a minimum. Dig holes big enough to ac-commodate the entire root system of the dormant vines you have pur-chased. Nurseries often prune dormant vine roots back to facilitate ship-ping, but this is not ideal for the growth of the vines. Those dormant roots contain carbohydrates needed for early-season growth, so refrain from further root pruning unless you find diseased or damaged roots.

When nursery stock have their roots jammed into a hole that is too

Holes are dug ahead of a crew planting green-grafted grapevines (grafted vines shipped and planted while actively growing).

small, the roots are bent back upward. This type of misplanting results in a phenomenon called "j-rooting," which occurs when roots do not grow normally in a horizontal direction away from the trunk but stay in a congestive mass at the planting site. J-rooted vines never perform well.

Keep the boxes of dormant vines closed and in the shade until those vines are needed. Take bundles from the box as they are needed by your planting crew, cut the string or strapping, and place them in five-gallon buckets full of water. Vines can remain in water for two to three hours with no damage; it is better that the roots remain saturated until they are planted. Do not leave dormant vines in buckets of water overnight or for any extended period of time. These are living plants, and the roots need access to both water *and* air to remain healthy.

Once the vines are planted, water them in by hand (one to two gallons per vine) to settle soil around the root system and to remove any air pockets that may dry out roots. Some growers use drip irrigation to accomplish this task, but many prefer the hands-on method so they can rest assured that the watering-in process has been accomplished correctly. If temperatures are still cool, there is reasonable ground moisture, and vines are fully dormant, deployment of irrigation tubing can be delayed a week or two. However, the irrigation system should be fully functional by the time growth begins in the spring.

Aftercare

The general rule for newly planted vines is to prune them back to two primary buds above the graft union. Pruning can be done the day of planting or any time up to the beginning of bud swell in the spring. Because parts of the root system are lost when vines are dug at the nursery and shipped, removing much of the dormant scion wood helps the vine regain equilibrium between the root system and the upper portion of the vine.

The use of grow tubes, which are plastic or waxed paper tubes placed over newly planted vines, is somewhat controversial. Clear plastic tubes create a greenhouse effect, leading to early, rapid growth of the vine inside the tube, but there is little evidence they really accelerate vine establishment. Either plastic tubes or cardboard grow enclosures provide some protection to young vines from contact herbicides used to control weeds in the new vineyard and can help reduce foraging by animals such as rabbits and deer.

In addition to the additional expense of grow tubes, the greatest drawback of using them involves the acclimation and hardiness of vines going into fall and winter. Because there is a green-

Traditionally, newly planted dormant vines are cut back to two buds after planting. Here, vines begin growth in early spring.

▲ To protect newly planted vines from animal damage and incidental herbicide contact, some growers use grow tubes on the vines. Tubes must be removed in late summer to avoid freeze injury in the fall and winter.

◄ Careful planning and follow-through can lead to successful vineyard establishment.

house effect, the temperature inside the tubes is warmer than outside temperatures during the day, but at night the tubes offer no insulation. The tubes can thus slow vine acclimation to winter conditions and exacerbate de-acclimation of vines in the winter, leading to severe winter injury in the first dormant season. If grow tubes are used, it is strongly recommended that they be removed during the first part of September so that vine acclimation and hardiness are not impaired.

CHAPTER 8

Vineyard Design

THERE ARE a number of logistical considerations to take into account when planning a vineyard for a particular location. Most decisions are made in relation to the specific dimensions of the parcel of land and the size of tractors and other implements that will be used in the vineyard. In many parts of the world, prime vineyard land comes at a premium, so high-density vineyards provide the greatest potential return per acre of land but may require specialized equipment.

Row Spacing

In both the eastern and western grape-growing regions of the United States the traditional spacing between rows was based on the space needed to take a team of horses and a plow down the row to eliminate competitive row center vegetation. Historically, California growers allowed ten to twelve feet between rows, while in New York nine feet was the norm. Today, vineyard row spacing ranges between seven and nine feet. Modern equipment is designed to fit these existing row dimensions, but in some parts of the world, narrow row spacing is more common and specialized equipment has been designed to operate in narrow row centers and to traverse steep slopes. The old working hypothesis is the 1:1 rule: a vineyard should have one foot of spacing between rows for every foot in trellis height to intercept adequate sunlight. Research in New York has shown that light interception can be optimized with rows spaced four and a half feet apart, but specialized equipment is needed to operate these vineyards and, under vigorous growing conditions, narrowly spaced rows can produce dense canopies that result in increased fungal disease pressure.

Ultimately, growers must choose how much land they are willing to dedicate to a given planting. Most growers strive to optimize yields on a per-acre basis, and spacing rows luxuriously far apart will reduce the

Rows are traditionally eight to ten feet apart. If rows are closer together, specialized equipment that is narrow enough to pass between the rows will be needed.

number of vines per acre, resulting in a reduced capacity to reach optimal yields. With equipment built to pass between narrowly spaced rows, some growers opt to space rows seven to eight feet apart. With standard equipment, nine to ten feet of row spacing should be the norm. Narrowly spaced rows require not only equipment designed to pass down the row but also increased vigilance on the part of the operators of equipment. In tight rows, equipment operators who take a quick look back at a spray pattern may find, upon facing forward again, that they are about to collide with a post.

Vine Spacing

Deciding how much space to leave between vines depends on the grape variety, rootstock, soil type, and anticipated rainfall at the vineyard site. Rootstocks can drastically impact overall vine vigor, and some varieties, such as 'Cabernet Sauvignon,' are inherently more vigorous than others. Likewise, a deep alluvial soil will promote greater vine size than either coarse sand or fractured limestone. While nitrogen greatly promotes increased vine vigor, water enhances grapevine growth better than any other substance. Vine size in higher rainfall zones will be intrinsi-

Spacing between vines in the row depends on soil depth and fertility as well as variety and rootstock selection.

cally greater than in locations that are more arid. Because of these variables, there is no one correct number of feet between vines in the row for every situation. The only constant is the principle that the greater the plant density, the higher the initial cost of establishment. Grafted grapevines are a significant investment, and decisions such as vine density will have long-lasting implications on the yield and fruit quality of a vineyard block.

While the standard spacing between vines has traditionally been eight feet, in the twenty-first century more and more growers have been opting to space vines closer together in the row. The reasoning behind the shift to planting vines closer together is somewhat complex. With increased plant density in the row, vines compete with each other for nutrients and light. The vines then exhibit less vigor than they would at lesser planting densities, but some viticulturists believe that the net yield remains the same, and because the crop load is spread across a greater number of vines, the fruit quality is better. In some locations in Europe, vines are planted in one-meter-by-one-meter planting schemes, with densities exceeding four thousand vines per acre. These vineyards for the most part are in cool climates, are planted on very limiting sites, and are tended by hand. For many vineyards in Texas, with deeper soils, higher temperatures, and more rainfall, the same densities would prove disastrous because there would not be enough space to accommodate the inherent capacity of the vines.

There is, of course, a point of diminishing returns on plant density. When growing conditions promote excessive vigor at a particular planting density, the result is that vines shade each other to the point where bud fruitfulness is reduced and vines get into a vegetative cycle. Planting vines closer than four feet apart in the row is probably not a good idea in any location in the southern United States, and on some sites, with certain varieties and rootstocks and certainly with high rainfall, greater distance between vines is probably a good idea. Visit vineyards in your area, talk to viticulture advisors or consultants, consider your budget, and determine what spacing and density are appropriate for your vineyard.

Row Orientation

In some parts of North America, such as the northeastern United States, the length of the growing season and limited light interception dictate the orientation of vineyard rows. In these situations, north/south rows maximize the amount of sunlight that is intercepted by the canopy. In areas with longer growing seasons, growers have greater latitude in deciding which direction to orient the rows in a vineyard. The single most important factor to consider is ease of using equipment in the vineyard. In a rectangular parcel of land, running the rows parallel to the length of the property is a much better choice than having short rows parallel to the short side of the rectangle. Turning from row to row takes time. While row length and orientation may not seem of much consequence in the planning stages, after a few hundred hours on your tractor and sprayer it will be easy to see why long rows are vastly superior to short rows.

Some growers opt to orient rows with the prevailing winds during the growing season. This orientation facilitates air movement throughout the vineyard, which means that the canopy will dry more quickly after a rain. Fungal diseases are generally driven by canopy wetness, so reduced drying time results in reduced disease pressure. If the vineyard is planted on a hillside, rows are commonly planted along the contours of the hill rather than up and down the hill. This orientation helps reduce erosion of bare soil under the vines and helps rainfall percolate into the soil rather than readily drain away from the vineyard and down the hill.

Both north/south rows as well as east/west rows have specific advantages in the hot, high-light intensity environment of Texas. Intense sun exposure from the south or west can result in sunburned fruit, so growers with rows that receive sunlight from either of those directions must take precautions. For example, although growers will typically remove leaves to reduce bunch rot as harvest time approaches, especially on varieties that produce tight clusters of fruit, they will leave much or all of the foliage in place on the side of a row that receives intense sunlight from the south or west. Growers can pull basal leaves on the eastern side of a north/south row or on the north side of an east/west row and still get the desired effect of reduced disease pressure without the risk of sunburned fruit.

Other Factors

Ease of equipment mobility through and around the vineyard is dictated by the amount of extra space allotted during the planning of the vineyard. The distance from the end post or anchor to a barrier such as a fence or ditch is called turn row space. With a tractor and sprayer, a minimum of thirty feet is needed to efficiently turn the equipment during vineyard operations.

If there is any possibility that you will employ mechanical harvesting, you must allow room to turn the equipment when planning the vineyard. While not quite as hard to maneuver as an aircraft carrier, a mechanical harvester cannot be turned around quickly or easily. Allow thirty to forty feet of turn row space if mechanical harvesting is being considered. Likewise, either placing alleys between blocks or dissecting large blocks can ease the movement of trucks and trailers into and out of the vineyard and thus facilitate harvest. Carrying picking trays by hand down several hundred feet of vineyard row will get old very quickly, so having a twenty-foot alleyway that provides passage for vehicles picking up fruit will aid in harvest efficiency.

Harvest is not the only time these passageways become useful. Proximity to supplies or water helps relieve the stress of vineyard work and will again become a factor influencing worker productivity. Vineyards should be designed to be efficient in production, and factors affecting the production of high-quality wine grapes are both viticultural and logistical.

The Growth Cycle and Grape Maturity

GRAPEVINE PHENOLOGY is the study and description of growth stages a vine goes through during the course of a growing season in response to environmental influences. Growers must be familiar with the basics of this phenology in order to make appropriate decisions about the management of the canopy and fruit load in a vineyard.

Unlike many other perennial fruit crops, grapevines bear fruit only on the vine growth of the current season. A single primary bud is capable of growing into a shoot that may bear one, two, or even three or more clusters of grapes. Fruitfulness of buds varies by variety and the physiological and environmental conditions to which the buds were exposed during bud initiation. Because of this extremely high reproductive capacity, grapevines can bear far more fruit than they can ripen. Thus, one of the annual tasks of the grower is to prune the dormant vines to reduce the fruit production potential of a vine, allowing it to grow a healthy canopy and properly ripen the fruit that remains on the vine.

Nodes and Compound Buds

A node is a part of a shoot or cane at the axil of a leaf, and it bears one or more buds. (Both growers and scientists commonly use the words *bud* and *node* interchangeably, so it can be confusing.) In many plants, there are single buds formed at each leaf axil, but grapevines typically have compound buds, which consist of three buds at each node.

During winter pruning, the grower retains the primary bud for fruit production. Below the primary bud is a secondary bud, which serves as a survival mechanism for the vine should the primary bud be killed by freeze, frost, or mechanical injury. The secondary bud usually bears one small cluster of fruit and can yield a partial crop after adverse winters or when the primary bud is lost. In some years, both the primary and secondary buds force together. Above the primary bud is a tertiary bud, which is a smaller secondary bud; it is often clusterless but can grow into a shoot if both the primary and secondary buds are killed. In the spring,

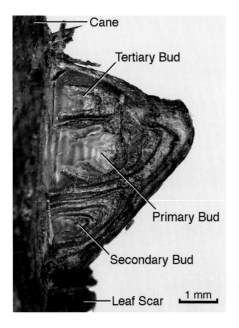

◀▲ Cross-section of a grape compound bud. (Photo by Martin Goffinet, New York State Agricultural Experiment Station)

▲ Dormant buds on a year-old cane. Grapes are borne only on current-season growth, and the number of nodes retained after dormant pruning will greatly affect vine yield.

◀ A fully swollen grape bud. This growth stage is sometimes referred to as doeskin.

these buds swell and begin to grow into shoots that are capable of growing as much as two inches per day.

Shoots and Flowers

As the bud continues to expand, it forms a shoot. Leaves grow along the shoot and, through photosynthesis, provide energy for

▶ Grape shoot of approximately one inch. Shoots grow and expand quickly in warm or hot weather.

▼ At five inches of shoot growth, flower clusters become clearly visible.

▼▶ At ten to twelve inches of shoot growth, flower clusters begin to separate and tendrils begin to attach themselves to trellis wires.

all vine functions. Along the shoot, typically opposite the second and third leaf, grape flowers develop. As the shoot continues to grow, the flower cluster expands until the initiation of bloom.

Many varieties begin to bloom when the tenth leaf on the shoot becomes flat. Shoots continue to expand during and after bloom, and tendrils (actually undifferentiated flower tissue) arise and grow at the leaf

Uniformity in shoot growth is essential to uniform bloom and fruit maturity.

Tendrils help actively growing vines to climb and attach to trellis wires.

axils past the last flower cluster. These tendrils are important tools in the ability of the vine to climb and attach itself to wires on the trellis. Shoot growth continues through the summer, and lateral shoots often develop from the axillary buds on rapidly growing shoots. As day length begins to decrease after the summer solstice, shoot growth slows and finally ceases.

Bud Initiation

It is important for all fruit growers to understand that, in any given year, they are growing two crops. The first and most obvious is the crop of fruit that is hanging on the vines that year. The second "crop" comprises the buds at the axils of the leaves in the renewal zone. The renewal zone is six to eight inches above the cordon, where the basal buds of shoots will develop for retained spurs, or short, woody vertical extensions of the cordon, and thus generate the crop to be grown in the subsequent year. For these buds to become fruitful, a vine must be properly managed to produce a crop of appropriate size and adequately supplied with water and nutrients. In addition, the canopy should be free of disease and exposed to sufficient sunlight. These conditions will optimize the fruitfulness of the buds on canes or spurs that will be retained after winter pruning.

Bloom and Fruit Set

Most grape varieties are hermaphroditic and thus contain both male and female flower parts. As shoots and flower clusters continue to expand, the calyptra or flower cap begins to separate from the grape flower. When the "cap" comes off, the flower pistil and anthers are exposed, and, with the drying of these tissues, the anthers dehisce and release pollen grains, which fall upon the receptive stigmatic surface of the pistil.

Bloom can take place over a couple of days to up to as much as two weeks, depending on weather and the uniformity of budbreak in a given year. Grapevines are wind pollinated and are prolific producers of pollen, so pollen availability is not a limiting factor to fruit set. Dioecious wild grape species may be insect pollinated, and grape flowers have a strong fragrance that attracts pollinators. Pollen grains germinate and grow down the pistil, where they encounter the egg; successful fertilization of the berry may or may not take place.

There are typically one hundred to two hundred florets or more on each blooming flower cluster, but only about 25 to 30 percent of those

Grape bloom occurs when the calyptra falls away from each floret, exposing the stigma and anthers. Grapevines have many more florets on a cluster than will become pollinated.

become fertilized. Those that are fertilized can potentially become grapes on the cluster, and, by definition, fruit set has occurred when young berries are approximately one-sixteenth of an inch in diameter. As the retained berries continue to expand they will go through a stage called shatter, in which some to many of the young berries abscise and fall to the

At shatter, unfertilized berries fall from the cluster and the fertilized berries continue to grow.
During the first thirty to forty days after bloom, grape berries grow very rapidly through cell division.

ground. This process can be alarming to some new growers, but it is normal. Appropriate fruit shatter levels can result in loose clusters that are far less prone to rot than tight clusters. Excessive shatter, however, can result in poor fruit set. It can be caused by boron or zinc deficiency, root loss, or adverse environmental conditions such as cold weather.

Berry Growth through Cell Division

For the first thirty to forty days after bloom, berries grow very rapidly if vines are healthy and supplied with adequate water and nutrients. This first period of rapid growth constitutes the time in which cell division is actively occurring within berries. During this initial

During the first thirty to forty days after bloom, grape berries grow very rapidly through cell division.

stage of berry growth, seed embryos begin to develop and organic acids and tannins begin to accumulate within the berry. Sugar content within the berries during this growth phase remains low.

Limitations, primarily suboptimal water availability in the growing season, can reduce the number of cells within a berry, which will ultimately reduce berry size. However, when cell numbers are lower and berry size decreases, the color and flavor of the wine may be improved because the skin/pulp ratio is increased. Because most of the flavor components in grapes and most of the color in red wine comes from the skins, having a higher proportion of skin to pulp results in greater color and varietal character in wines. In Texas, however, rains that fall later on in the growing season can cause vines to take up high volumes of water. When the water reaches the berries, splitting commonly occurs, especially if berry cell numbers were intentionally reduced by withholding water during the first month after bloom. If grapevines are carrying an excessive crop, fruit can be removed in the first thirty days after bloom, which can result in an increase in the quality of the fruit that remains.

Lag Phase

After cell division has ceased, berries go through a developmental process known as lag phase. During lag phase, berries do not appear to

During lag phase, visible growth slows greatly, but seed maturation and berry acid formation are under way.

be growing or changing much, but seed development continues, and organic acids, primarily tartaric acid and malic acid, are being produced. Berries remain firm, but berry chlorophyll content begins to degrade.

Lag phase usually lasts between three and four weeks, depending on the variety and growing region. By lag phase, the canopy should be fully developed and capable of conducting sufficient photosynthesis to maintain healthy vine function, ripen the crop, and begin to store carbohydrates for later use. Irrigation scheduling, nitrogen fertilization, and canopy hedging are all tools growers can use to maintain a healthy canopy while it assists the vine in using the products of photosynthesis to help ripen crop.

Véraison and Ripening

Véraison is the term used to describe the onset of ripening in grapevines. In red- or black-fruited varieties, véraison is signaled when 5 percent of the berries show first color and, in white- or green-fruited varieties, when 10 percent of the fruit has changed from an opaque color to translucent. Véraison is also signaled when the soluble solid content of the fruit reaches 7 to 8 percent.

Ideally, at this point the canopy has significantly slowed or stopped new shoot growth. Shoot tips should cease growth, while maintenance of a healthy, viable canopy remains a priority. Additional canopy hedging may be needed to slow vegetative growth. With hedging, especially in wet years, lateral shoot formation is increased and vertical hedging to remove lateral shoots may be required. Remember, pruning green tissue during the growing season is a dwarfing action and can be easily overdone, but on very vigorous sites hedging may trigger an increase in lateral budbreak, further exacerbating the problem.

At the onset of véraison, fruit soluble solid content can commonly increase 2 percent per week, but the rate begins to slow considerably as maturity approaches. Rain events of one or two inches can easily roll back sugar development by 2 percent or more immediately following the rain, but the rate will pick up as the vineyard begins to dry out. Supplemental irrigation is used to keep vines photosynthetically active, but it should not be excessive.

Monitoring Grape Maturity

Perhaps the most contentious issue between grape grower and winemaker is deciding when to harvest. Growers generally are paid upon delivering as many tons of fruit as they can grow while achieving a specific

Véraison is the beginning of fruit maturation, when white-fruited varieties become translucent and red-fruited varieties begin to color.

target, usually a specified soluble solids level. In an ideal world, growers would always make a decent profit on a block of grapes and winemakers would always have the high fruit quality they want. In reality, grape contracts are usually written with some measurable parameter as a benchmark for "ripeness."

While sugar content, usually expressed as degrees Brix (total fermentable content, including some proteins and amino acids), is an easy way to monitor grape quality development, it is by no means the most expressive of actual grape quality. As sugar content in fruit rises due to photosynthesis, organic acid content, necessary for desirable flavor components, and pH stability both fall, especially during hot nighttime temperatures. The best winemakers use degrees Brix, titratable acidity, pH, and, perhaps most importantly, flavor to make the decision on when to harvest a crop.

Wood Maturation

By the time the fruit starts to mature, vegetative shoot growth should have slowed or stopped and maturation of green shoots into woody canes has begun and can continue as long as there is a functional canopy. These

Mature fruit just before harvest.

two processes continue simultaneously, but at this stage of growth, fruit maturation is usually the primary photosynthetic sink, which refers to the location to which carbohydrates are preferentially delivered to complete the plant growth or maturation processes. Thus, when a vine has achieved 100 percent of its production potential, there are few photosynthates left to ripen canes that will yield new growth and a crop for the following year.

In many locations, a long period of post-harvest photosynthesis can solve this problem and provide enough photosynthates to ripen wood after the crop has been harvested. In some years, especially in the High Plains, harvest may be relatively close to first frost, and thus the period of post-harvest photosynthesis may be very short. In all regions of the state, if the canopy is not kept healthy, with adequate water and nutrients, and kept free from insects and diseases, a canopy can quickly senesce, resulting in premature leaf fall. Without foliage, photosynthesis cannot occur, shoots cannot mature into woody canes, and vines cannot store carbohydrates to increase winter hardiness. Wood that has not formed periderm (woody tissue formed during the maturation process of annual shoots) will be killed with the first hard freeze.

Late Fall and Winter Growth

In the Texas climate, where the soil does not freeze during winter, fall, and early winter is an extremely important time for new root growth. This growth takes place only when there is sufficient water, so in years with dry winters, some irrigation may be essential. In the fall, when roots are actively growing, they are foraging for and absorbing nutrients, just as they are during the spring and summer. Nitrogen, zinc, and boron are taken up during the fall and winter for use in the growing season that follows.

Grapevine Physiology

W HILE IT IS not the intent of this text to offer in-depth explanations of the complexities of grapevine physiology, it is important for growers to understand a few basic principles that drive grapevine growth and fruit maturity. Unlike annual crops such as beans or squash, which are replanted each year, grapevines, like all perennial crops, require a long-term strategy that accommodates every season and each corresponding condition of the vines. Failure to understand and to respond to critical threats typically results in vines that are predisposed to environmental stresses. Once vines are severely compromised by conditions such as winter injury, they may be saved and retrained, but secondary problems that limit the life and productivity of a vineyard commonly arise. Growers must therefore follow a long-term strategy to maintain a healthy and productive vineyard.

Photosynthesis

The driving force that powers growth and reproduction of all green plants is photosynthesis. By harnessing the energy of the sun, green plants use atmospheric carbon dioxide (CO_2) and water (H_2O) to produce carbohydrates (CH_2O) and return oxygen (O_2) to the atmosphere. While this explanation of the process is greatly simplified, it is enough to describe how carbohydrates are produced in green plants. Photosynthesis is carried out by chlorophyll contained within leaf tissue. Chlorophyll appears green to the human eye because red- and blue-spectrum light rays are absorbed while green-spectrum rays are reflected. For plants to be photosynthetically efficient, they must have ample light exposure, sufficient water, moderate temperatures, freedom from damaging insects and pathogens, a sufficient supply of essential nutrients, and an appropriate

Photosynthesis is the sole energy source for the vine. For vines to be photosynthetically efficient, they must have adequate supplies of water, nutrients, and sunlight and freedom from insect and disease injury.

atmosphere, devoid of chemical pollutants. If any of these conditions are insufficient, photosynthetic capacity will decrease.

While in some northern US grape-growing areas sunlight may be limited because of persistent cloud cover, in Texas cloudy conditions rarely cause problems. Instead, light limitation typically occurs only when a grapevine is shading itself. To remedy this problem, canopies should be managed to ensure that basal leaves that subtend (are opposite from) grape clusters are well exposed to light. Although there are shoot-positioning machines that are capable of arranging canopies for maximum performance, they are typically cost prohibitive for small growers, who must rely on hand labor to accomplish this task.

Grape buds on shoots of the current season form at or shortly after bloom, and shoots that are shaded at this point in the growing season will produce buds that yield lower fruitfulness in the next crop. As fruit begins to ripen it is also important for the leaves that subtend clusters and leaves on the lower two feet of the canopy to remain properly exposed to light so that they can efficiently photosynthesize and achieve optimal sugar accumulation in the clusters. As with all perennial crops, the key to keeping plants healthy and producing high-quality fruit is successful interception of sunlight.

Plants will die if subjected to specific high and low temperatures. Between those two lethal temperature extremes, plants will survive, grow slowly, or grow optimally. Numerous studies on grapevine photosynthesis suggest that grapevines have optimal photosynthetic capacity when ambient air temperature is between 77 and 86 degrees Fahrenheit. Because fruit ripens in Texas during July, August, and September, very high summer daytime temperatures can become a limiting factor to building soluble solids in the fruit. Little can be done to alter the temperatures Texas experiences during the summer, but keeping vines well supplied with water and nutrients and eliminating competitive vegetation can moderate the effects of extreme summer heat. Choosing varieties that ripen well in this climate is also critical to the long-term success of a vineyard operation.

Most living organisms are composed of 70 to 90 percent water. In plants, water is essential not only as a constituent element but as a necessary partner in photosynthesis, as well as all other physiological processes within the plant. Water and carbon dioxide are the primary components of the carbohydrates created by photosynthesis, but the importance of water to the plant is far more complex. While plants absorb water through the roots, they also emit water through gas exchange pores called stomata, located on the lower side of leaves. With proper water availability, these stomata open fully to allow oxygen and carbon dioxide into the leaf while emitting water vapor back into the atmosphere. This release of water allows it to evaporate, which produces an endothermic condition, thus cooling the leaf itself. This cooling effect allows a plant to continue photosynthesizing when ambient temperatures exceed the optimal range.

So-called experts have been known to recommend that Texas growers withhold water during the period from véraison to harvest. The reason these advisors give is that the water deficit concentrates sugars in the berries during the maturation process. That approach may work in more moderate climates, but in the heat of Texas summers, withholding water at any time to stress vines artificially will stop photosynthesis and produce effects ranging from counterproductive to catastrophic. Sugars accumulate through the active process of photosynthesis. While it is true that excessive moisture that leads to runaway vegetative growth should be avoided, vines need to be supplied with sufficient water throughout the growing season to maximize photosynthetic efficiency.

An adequate supply of essential elements is necessary for grapevines to produce healthy plant tissue, to conduct physiological processes, and to regulate a multitude of hormonal interactions. Vines that are not adequately supplied with macro- and micronutrients struggle to grow and produce high-quality fruit. Nutrient deficiencies, especially of nitrogen, iron, zinc, and magnesium, can degrade chlorophyll, reducing the ability of leaves to photosynthesize and commonly resulting in a premature senescence of foliage. Analyses of soil and petioles can help predict the nutrient needs of a vineyard, but a grower needs to have a thorough understanding of how nutrients move in soils and plants in order to react appropriately to changing conditions in the vineyard.

It is also important for growers to understand that both insects and diseases can directly affect photosynthesis. Insects that feed on foliage or plant pathogens such as powdery mildew or downy mildew can seriously reduce the amount or health of the canopy of a grapevine and negatively affect growth and fruit production.

The Carbohydrate Bank

Growers could benefit by thinking of grapevines as having a sort of bank account. The currency in that bank is carbohydrates, and the goal of growers is to keep the account as full of carbohydrates as possible. Carbohydrates are produced in only one way, and that is photosynthesis, hence the importance of understanding factors that limit that vital process.

Carbohydrates are expended in a number of ways over the course of a growing season, and the destinations of those expenditures are called sinks. Initial growth in the spring is driven by the use of carbohydrates that have been stored in the vine (primarily the root system) since the previous season. Early in the spring season, the primary photosynthetic sinks are the growing shoot tips and the expansion and maturation of foliage. As the canopy develops and leaves mature, they become the source of additional carbohydrates. The additional carbohydrates continue to fuel the rapid development of the canopy, which physiologists call the period of grand growth. After flowering and fruit set, the primary sink of a grapevine then becomes the fruit crop. As the crop develops, vegetative growth starts to slow. If vines are overcropped, vegetative growth can drastically decrease, severely reducing the supply of carbohydrates in the fruit load. In addition, the diminished canopy produces fewer carbohydrates. In this scenario, vines have started to become overdrawn at the bank and the prospect of a properly ripening crop appears to dim.

Ripening of fruit is the primary way carbohydrates are depleted from grapevines. To avoid depleting the carbohydrate account through over-cropping, which yields poor-quality fruit, the grower should thin the crop within the first twenty-five to thirty days after bloom. The vines then respond by increasing the quality of the remaining crop, thus balancing the system.

After harvest, the next active photosynthetic sink is the ripening of green annual shoots into woody canes. Plant physiologists refer to this filling of vines with late-season energy as carbohydrate loading. The transformation of green tissue into woody brown tissue is called periderm for-

With an adequate carbohydrate supply, green shoots begin to form periderm and ripen into canes capable of surviving the winter.

mation. Wherever periderm has formed in the canes, those portions will survive freezing temperatures in winter. Periderm forms from the basal portion of the shoot toward the apical tip. Apical portions of the shoot that fail to develop periderm will die with the first hard winter freeze.

Even after harvest, it is vitally important for vines to be well cared for to maintain the health of the vine for the season to come. Adequate moisture levels, weed control, insect and fungal pest control, and ample nutrition must all be maintained well into the fall to keep vines from prematurely defoliating in late summer and fall. There are not many blessings of trying to grow grapes in the Texas climate, but the considerable length of the growing season can allow for significant post-harvest photosynthesis. Because, during the active growth period, roots are fourth in line to receive resources, this period of late-season photosynthesis is critical for replenishing carbohydrate reserves in roots and for promoting important dormant-season root growth.

Dormancy

Grapevines are indeed a temperate fruit crop, and outside of tropical areas vines respond to a variety of environmental cues to control the annual growth cycle. Vine growth is initiated in the spring, when winter dormancy has been overcome and the vines have been exposed to sufficient heat. In midseason, after the passing of the summer solstice, day length begins to shorten and vines respond by slowing vegetative growth and then maturing green shoots into canes that can survive the winter. In early autumn, vines begin to harden off, the canopy naturally begins to senesce, and cool temperatures begin to prepare vines for winter. During the ideal temperatures and moisture levels Texas vineyards may experience during the fall, the buds that were formed at the axils of the leaves during the growing season are suppressed from growing by endogenous growth inhibitors within the buds. Abscisic acid (ABA) is the primary growth inhibitor, but it is believed that other compounds are also responsible for regulating dormancy.

Although the dormancy mechanism in grapevines may be relatively weak compared to that in other temperate fruit crops, such as apples, it does exist, and vines need to survive through the winter in order to be able to grow again in the spring. The growth inhibitors are broken down by cold winter weather, but subfreezing temperatures actually do little to overcome dormancy; temperatures ranging from 38 to 48 degrees Fahrenheit are the most efficient temperatures for breaking down growth inhibitors.

When horticulturists measure the chilling a given location receives, they measure ambient air temperature. However, internal bud temperatures are more indicative of how vines are actually responding to the environment, and research shows that cold, rainy days appear to be the best conditions to assist vines in overcoming dormancy. For reasons not thoroughly understood, different varieties of grapevines appear to have different chilling requirements based on the origin of the parent species. For example, *V. vinifera,* primarily native to the Middle East and southern Europe, is adapted to shorter, milder winters. Conversely, *V. labrusca,* native to the northeastern United States, has a much more substantial chilling requirement. Once the chilling requirement of a given variety has been met, the vine begins to grow. As mentioned, the dormancy mechanism of grapevines is relatively weak, so even after warm winters vines will eventually wake up and grow. The benefit of a good, cold winter, however, is that budbreak following such a winter is uniform rather than protracted, which results in a crop that is uniform in growth and ripens evenly. Growth regulators that mimic naturally occurring compounds produced by plants are used in some parts of the world to help overcome dormancy, but the use of these materials is generally confined to tropical and subtropical latitudes, where vines are commonly cropped twice per year.

Hardiness

While hardiness is somewhat related to dormancy, this phenomenon is a unique process that growers need to fully understand in order to make management decisions that will keep vines healthy and productive. As vines prepare for winter, environmental cues such as shortening photoperiod, cooler temperatures, and finally the first frost take vines to the doorstep of winter. As temperatures get colder, winter, vines become more and more hardy until they reach their maximum hardiness level.

Grape varieties vary considerably in their inherent ability to withstand winter temperature extremes, but environmental conditions themselves play a huge role in the ability of a vine to avoid freeze injury. The general rule of thumb is that vines gain hardiness very slowly but lose it quite rapidly. For example, when winters are cold and stay cold, most temperate grape varieties can withstand temperatures below 5 degrees Fahrenheit. The major problem encountered by grape growers in Texas is the loss of hardiness caused by midwinter warming spells. When

Vines can exhibit winter injury in the spring, when dead tissue is immediately recognized, or vines can collapse in summer with the added stress of high water demand.

temperatures warm to between 70 and 80 degrees Fahrenheit or more, followed by a passing cold front that drastically drops the temperatures, vines can be damaged or even killed by cold injury. The winter of 2010 in the Texas Hill Country saw temperatures drop steadily from freezing and finally hit a low of 7 degrees Fahrenheit. However, there was no trunk, cordon, spur, or bud injury in any area vineyard because vines were well acclimated to winter freezing conditions. More typical of winter injury episodes in Texas was the winter of 2011, in which temperatures warmed in late January, reaching 74 degrees on February 1. Following passage of a cold front, temperatures on February 2 dropped to 11 degrees Fahrenheit. Area vineyards suffered widespread bud, cordon, and trunk loss, and in some areas widespread vine death was reported.

Rapid, dramatic shifts in weather conditions are a major challenge of growing grapes in Texas. The freeze injury in 2011 was greatly exacerbated by a severe drought that had been ongoing for six months before this episode. Vines that are under drought stress are much more susceptible to winter injury than those that are well supplied with water. Although many growers do not normally irrigate much, if at all, during the winter, under severe drought conditions thoroughly watering a vineyard for a few days before a freeze event can greatly reduce vine injury.

Although we cannot do anything about the weather, growers can take steps to minimize vine injury even during events like the winter of 2011. The answer is carbohydrate loading. Carbohydrates are not only the energy source used by grapevines; they are also the antifreeze mechanism. Carbohydrates are complex sugars. At 32 degrees Fahrenheit, a glass of water will begin to form ice crystals. If a teaspoon of sugar is dissolved in that glass of water, the freezing point is lowered. The more sugar that is dissolved in that water, to the point of saturation (at which no more sugar can go into solution), the greater the suppression of the freezing point. Growers can do the same thing with grapevines. It is not uncommon to see young vineyards with a healthy canopy go through a freeze in late November, when temperatures fall into the mid-twenties, and not suffer any frozen leaves. Why? Because a young vineyard that has not expended a majority of its carbohydrates bearing a crop has foliage that is full of carbohydrates, and the intercellular fluids will thus resist freezing well below 32 degrees Fahrenheit.

By maximizing post-harvest photosynthesis and carbohydrate loading to the greatest extent possible, growers have done everything they can do to help a grapevine resist winter injury. By keeping a vine well supplied with

water as it enters a cold event, a grower enables the mobilization of carbohydrates within the vine and heightens the cold resistance of those vines.

Root Growth and Function

While the parts of a grapevine below the soil surface are not typically
visible, the root system is nonetheless vital to the health of the vine. The
root system has a number of functions, some quite evident and others
that are a little less obvious.

First and foremost, grapevine roots provide anchorage; they are what
physically hold vines in the ground. Rootstocks are derived from several
Vitis species, and many of them have distinctly different types of root
systems. Some are deep-rooted, others shallow; some can tolerate high
soil moisture, while others need well-drained soils. Matching the type of
root system to the specific vineyard site is thus a high priority.

The second rather obvious function of roots, and the most important
one, is to absorb water and nutrients. It should be noted that absorption
of both water and nutrients mostly occurs in new root tips, which means
that if roots are not actively growing, there is little absorption of water
and nutrients. Because of dry summer growing conditions in Texas, actively growing roots are typically concentrated immediately under the
emitters of a drip irrigation system. When there is adequate soil moisture, the entire vineyard floor is covered with an active, functional root
system, and grapevines thrive in those conditions. While drip irrigation
can supply water to a grapevine, and that water can be moved within
the vine, extended drought can cause much of the root system in mature
grapevines to be marginally functional. Drought conditions will not necessarily kill the roots, which are supported and kept alive by water picked
up under the drip zone, but without adequate water in the immediate
vicinity of roots, they will lie quiescent and do little to actively support
vine growth.

Because roots act as an important storage site or bank for carbohydrate
reserves, it is extremely important to remember that excessive cultivation
to control competitive weeds on the vineyard floor ultimately results in
the destruction of grapevine roots and the loss of important carbohydrate
reserves. While root morphology differs among rootstocks according to
their genetic parentage, as a rule of thumb, 80 percent of the root system
of most grapevines is located in the top six to eighteen inches of soil (depending on soil texture). Certainly, in deep, coarse, sandy soils, functional
rooting depth is greater, but it is important to remember that roots need

both oxygen and available water to absorb water and nutrients. In heavier soils, the functional rooting depth may be less than six inches. If growers are looking to devigorate a vineyard, this is an appropriate cultural practice. If they are simply seeking to manage competitive vegetation, there are far less destructive practices that accomplish this objective. A disk may be an appropriate tool to prepare a site for planting, but it is not really an appropriate tool for managing vineyard floors.

Roots are also an important source of plant growth regulators. Roots that are under moderate water stress can also produce abscisic acid during the growing season, which can enhance fruit ripening. Earlier in the season, before bloom, new root tips are responsible for the production of cytokynins, which regulate fruiting and vegetative growth patterns. There are documented cases in which deep cultivation is thought to have caused economically important crop losses due to the reduction of cytokynin production before bloom, thus triggering a significant abortion of fruit.

During certain times of the year, grapevine roots exhibit an active flush of growth. The first period is shortly after bloom. While the developing cluster is a prime photosynthetic sink, the development of a full canopy provides sufficient energy for some of the carbohydrates to be diverted to the growth of new root tips. This root growth is important because vines at this point are starting to demand higher levels of both water and nutrients. During the post-bloom period, roots are actively absorbing nitrogen and potassium. Research proves this activity because dry post-bloom periods commonly result in low potassium levels in petiole samples, whereas wet springs yield high potassium uptake by vines.

The second and most enduring phase of root growth in Texas vineyards (where the soil does not normally freeze) is during fall and winter. Grapevines actively forage for and store specific plant nutrients that are important for early-season growth and for flowering and fruit set. Dry fall and winter periods limit this activity and result in vines that are slow to come out of dormancy. Although roots generally remain out of sight, understanding and caring for the root system of grapevines is every bit as important as the treatment of the aboveground portions.

Rootstock Selection

ROOTSTOCKS are genetically distinct cultivated varieties of grapevines used to induce or reduce scion vigor or to overcome specific soil limitations. While a particular grape variety may ripen well in a given location, the characteristics of that vine, when grown on its own roots, may not match up with the restrictions of a particular climate, soil type, or other site limitation. Grafting, a technique that joins two varieties or species of plants together, has been used for thousands of years, and it is the reason growers are able to choose a specific type of vine suited to soil conditions and another distinct scion or fruiting variety suited to other environmental conditions.

The inadvertent movement of phylloxera from the New World to the Old in the 1860s necessitated the use of resistant rootstocks to overcome this root-feeding insect pest, and using grafted vines is now common in most grape-growing areas of the world. T. V. Munson, who was a horticulturist and grape breeder in Denison, Texas, is widely credited with saving the French (and most of the Old World) wine industry because of his introduction of rootstocks using native Texas grape varieties that had tolerance to phylloxera and limestone soils.

Most rootstocks in use today bear strong resemblance to their wild parents, and many of those varieties are pistilate (having only the male flower) and thus bear no fruit. There is no perfect rootstock, but understanding the strengths and weaknesses of various rootstocks among the broad selection that are commercially available can help a prospective grower best match the rootstock with specific scions and the physical site to produce a long-

Grafting machines, such as this omega bench grafter, are used to join rootstocks with the bud of a fruiting or scion variety.

◄ After spending a few weeks in a dark, warm, humid chamber, plants have grown together and are placed in pots or nursery rows for initial growth.

► Here, vines for experimental research plots are being grown in the greenhouse before being transplanted to the field in the summer.

lived, healthy, productive vineyard. It is a general rule of thumb that when breeding plants for resistance to specific threats, the breeder looks to plant material that is native to the area where the threats exist.

Own-Rooted Vines

Grapevines throughout the ages were most commonly propagated as own-rooted, or self-rooted, vines. Cuttings made from year-old grape-vine canes root readily, so a single vigorous vine can provide enough plant material to propagate more than a hundred new plants each year. In some parts of the world, including the United States, the use of own-rooted vines is still commonplace.

Where there are no threats presented by soil limitations, insects, or pathogens, using inexpensive, true-to-type plant material makes good viticultural and economic sense. *Vitis ×labruscana* varieties, such as 'Concord' and 'Niagara,' are commonly grown on their own roots in the Lake Erie region of New York and in the Pacific Northwest. Inexpensive own-rooted vines can be purchased from nurseries, and if something catastrophic happens to the trunk, such as mechanical or severe freeze injury, shoots arising from the root system will bear desirable fruit.

There are, however, numerous reasons to consider purchasing grafted vines. Biotic threats include phylloxera, several species of nematodes, and soil-borne pathogens, including post oak root rot (*Armellaria mellea*) and cotton root rot (*Phymatotrichopsis omnivora*), both of which are common in much of Texas and the southern United States. All of these represent serious threats that can devastate a vineyard in a short period of time, and the susceptibility of own-rooted vines to these factors should not be ignored. Given the other expenses involved in vineyard establishment, choosing own-rooted vines for economical reasons may be short-sighted.

Using and Choosing a Rootstock

Given that nematodes, phylloxera, cotton root rot, Pierce's disease, and numerous soil-borne pathogens are endemic to the native range of *Vitis berlandieri* and other native Texas *Vitis* species, it is no coincidence that these species are parents to many of the common rootstocks used around the world today. Appropriate rootstocks can also help vineyards overcome physical and chemical soil restrictions, including poorly drained sites, high calcium bicarbonate levels, and high sodium levels. Because all of the aforementioned insects and diseases are native to Texas and much of the southern United States and because native grapevine species have coevolved with these threats, native grape varieties have, over time, become resistant to or tolerant of these limitations.

One of the most important situations for the new grower to avoid is buying new vines on a rootstock that is not optimal for your site. Nurseries are in the business of selling vines, and last-minute orders may put you

Rootstocks can play a big role in imparting or controlling the vigor of a scion. Although seemingly mismatched in size, this rootstock is accomplishing the goal of managing the vegetative vigor of this vine.

in touch with a sales representative who may not offer you the best advice on rootstock selection. The vines they currently have in inventory may not be what you need, but it is what they have to sell. Contact your nursery fifteen months in advance of anticipated planting date. For numerous reasons, this much time is needed to make other planting preparations, so make informed decisions on variety and rootstock early in the planning process and find a reputable nursery that can propagate exactly what you need.

Native Texas Rootstock Parents

Texas is blessed with a wide variety of native *Vitis* species that have adapted to every corner of this diverse state. From the high-rainfall, acidic soil sites in East Texas to the arid, calcareous soils of the Hill Country and West Texas, the exact number of grape "species" remains to this day a topic of debate among viticulturists, botanists, and plant systematists. The fact that there is so much variation among grapevines native to the state ultimately works well for Texas growers in that these native vines have endured centuries of climatic diversity and soil limitations. The state is rich with natural *Vitis* resources, and growers' success is tied directly to that treasure. However, only two species, *V. riparia* and *V. rupestris,* root well.

Vitis riparia

Many of the plant selections from *Vitis riparia* used for rootstock breeding came from high-rainfall, acid soil locations throughout the Midwest and the northeastern United States. Rootstocks with this parentage are able to tolerate "wet feet" (waterlogged soil), accelerate scion maturity, and maintain moderate scion vigor under high-rainfall conditions. Phylloxera is native to much of the native range of *V. riparia,* so tolerance to that insect is inherited with the use of this parentage in rootstocks.

Cold hardiness is also a characteristic of this variety, but, unfortunately, many *V. riparia* hybrids break dormancy relatively early, leading to susceptibility to spring frost. This characteristic is more problematic in scions than rootstocks with this parentage, but it is the nature of this species because it comes from areas with relatively short growing seasons.

Major drawbacks of this species include intolerance to high-pH soils and drought, as well as variable nematode resistance.

Vitis rupestris

Although native to North America, *Vitis rupestris* is an endangered species among grapevines because it is a favored food source for cattle, deer, and other browsing animals. Growing wild in creekbeds and dry streams, this species is deep rooted and can thus hold itself in place in very erosive environments. This feature allows it to tap into water tables, but it is consequently sensitive to drought under most cultivation conditions. With irrigation, stocks with this parentage have relatively high vigor, which may be an advantage in shallow soils. Tolerance to phylloxera is acceptable, but nematode tolerance may be variable depending on the nematode species. *V. rupestris* does not like poorly drained soils and has only moderate tolerance to high soil calcium content.

Vitis berlandieri

Vitis berlandieri is the grape that saved the wine industry of Europe. For that alone, Texas grape growers should know that the "native children" have had a major impact on the history of the world. Although the phylloxera pest is native to North America, infestation is not particularly common in Central Texas because native grapevines are resistant, which results in low insect populations.

V. berlandieri is also very tolerant of relatively shallow, high-lime soils, expresses moderate vigor in scions, and has intermediate tolerance of nematodes and wet feet. Of the modern stocks currently available at the writing of this manuscript, few without *V. berlandieri* parentage should be considered for most growing regions of Texas.

Vitis champini

While some viticulturists, including T. V. Munson, considered *Vitis champini* to be a distinct species, modern plant systematists believe it to be a natural hybrid between *V. rupestris* and *V. candicans* (*mustangensis*). Rootstocks with this parentage appear to have high tolerance to soil-borne pathogens, but they appear to be only somewhat resistant to root knot nematodes (*Meloidogyne incognita*), common on sandy soils across the South. Good PD resistance, relatively high resistance to cotton root rot, moderate vigor, drought resistance, and moderate tolerance of high calcium `content soils, this parent is adapted to all but sandy sites with extremely high root knot nematode pressure.

Table 11.1. Rotstocks commonly encountered in the modern commercial Texas grape industry

Rootstock	Parentage	Phylloxera	Root Knot nematode resistance	Dagger nematode resistance	Drought tolerance	Tolerance to poorly drained soils
SO 4	Vitis berlandieri ×riparia	H	M	M	L	M
5 BB	Vitis berlandieri ×riparia	H	M–H	M	M	L
5 C	berlandieri ×riparia	H	M	L	L	M
R 110	Vitis berlandieri × rupestris	H	M	L	H	L
1103 P	Vitis berlandieri ×rupestris	H	M	L	M	M
Freedom	C 1613 × Vitis champini	L	H	H	M	L
Harmony	C 1613 × Vitis champini	L	H	L	M	L
Salt Creek (Ramsey)	Vitis champini	H	H	H	H	M
Dogridge	Vitis champini × V. candicans?	M	H	?	M	M
Champanel	Vitis champini × 'Worden'	M	L	L	M	L
Riparia Gloire	Vitis riparia	H	M	M	L	H
101-14 Mgt.	Vitis riparia × rupestris	H	M	M	L	M
3309	Vitis riparia × V. rupestris	H	L	L	L	M

Salinity tolerance	Tolerance to free lime in soils	Vigor	Soil adaptation	Resistance to cotton root rot	Resistance to Pierce's disease	Other comments
L	M	M	Loams, clays, sandy clays	L	M?	Some confusion with 5C
M	M	M	Sandy clays, loams, clays	M	L	More drought tolerant than 5C
M	M	M	Finely textured soils	L	M	Apparently slightly more vigorous than SO 4
M	M	H	Coarse soil types (sand, gravel)	L	M	Excessively vigorous under high rainfall
M	M–H	M–H	Widely adapted	M–H	H	Popular, versatile stock except with high vigor scions
M	M	H	Sandy, well-drained soils	?	L	Excessively vigorous in fertile sites
M	M	M	Coarse, well-drained soils	?	L	Similar to Freedom, but slightly less vigorous
H	M	VH	Light, sandy, infertile soils	?	H	Suited to coarse, unfertile soils
L	M	VH	Sands, clay loam, neutral to mildly alkaline soils	H	H	Hard to root and graft; reports of excessive vigor and suckering
L	M	M	Tolerates all soil types	H	H	Not widely available
L	L	VL	Very low vigor stock for acid soils	?	?	Prefers deep fertile soils
L	L	M	Sands to clays; prefers neutral pH soils	?	?	Popular stock for near-neutral pH soils
L	L	M	Prefers deep, well-drained soils	L	?	Cold hardy stock for acid, sandy sites with low nematode pressure

Vitis monticola

Little used in current rootstocks, *Vitis monticola* is a low vigor, native Texas grape species found naturally in shallow, calcareous soils of the Texas Hill Country. Drought tolerance, very low vigor, and resistance to PD and soil-borne pathogens may make this native species important in the newly bred rootstocks of the future, but it is very difficult to propagate.

Vitis vinifera

Most of the desirable fruiting varieties of grapevines in the world today have their origin in *Vitis vinifera.* Fruit from this species exhibits exceptional quality, but other than tolerance to high-pH soils, *V. vinifera* has little to contribute as a parent in rootstocks. Phylloxera susceptibility is the Achilles' heel of this species as a rootstock parent.

Other Species

There are numerous other species of grapevines native to the United States that either now have or will have a role to play in the development of new rootstocks for specific areas of the United States. Adaptability to a wide variety of soil and climatic conditions, as well as to various soil-borne challenges, provides breeders with the genetic resources they need to address the threats the grape-growing industries will face. Other Texas species include *V. rotundifolia, V. rufotomentosa, V. arizonica, V. candicans, V. doaniana, V. acerifolia,* and others. The natural diversity of native *Vitis* resources within the state remains one of the greatest treasures of Texas and a source of optimism for the future of the industry and of growers' ability to overcome unforeseen problems.

Commercially Available Rootstocks

The accompanying chart lists rootstocks commonly encountered in the modern commercial Texas grape industry along with the characteristics of each. The chart also lists some heirloom varieties that may be of value in specific situations.

This list is not all-inclusive, and research is currently under way to evaluate newly available rootstocks and to more accurately define differences between existing available rootstocks. Rootstock selection should be made on the ability of a stock to overcome limitations of a given site and to provide the appropriate amount of vigor when receiving a specific scion graft. Talk to other growers in the region where you intend to work to find out what works and what does not and to understand the reasons behind the good or bad performance of the various rootstocks.

Selecting Fruiting Varieties

A FTER SELECTING an appropriate site and rootstock, new commercial growers must make yet another decision with long-term consequences. Choosing which varieties of grapes to grow is a decision that should be carefully considered before placing an order with a nursery. Commercial demand by wineries, intrinsic risk of growing specific varieties, and availability from nurseries will all play a role in when and what you plant.

New commercial growers should avoid the temptation of trying many varieties, planting small sections of them, and then waiting to see which works best. The blocks you plant should be of commercial size. If you are planting an acre, choose one, maybe two varieties tops, and tend it well. If you have only one hundred vines of a variety, no winery will want to see you show up with just five hundred pounds of fruit at their crush pad. They may work with you if you can deliver two tons of fruit, but, by all means, *have your market established before you plant a single vine.*

If you are growing for a specific winery, that operation will of course want you to grow specific varieties. Before jumping into these varietal decisions, carefully consider the difficulty in growing a particular variety and who will be bearing the risk of this choice. Losing a crop to spring frost or disease pressure or failure to ripen to industry standards will be your loss, not that of the winery. If the wineries themselves are growing that variety in your area, it is a good sign that the managers are invested in the operation. Such is not always the case, but new growers need to thoroughly research the risks and rewards of a given varietal choice before making that long-term investment.

Texas is a very large and diverse state, with many different growing regions. While a given variety may be grown in much of the state, fruit quality may be markedly better in certain regions. Because of the complexity of possible growing scenarios, this discussion of varieties can in no measure be complete. There are some varieties not mentioned here that are being grown and made into wine for commercial sale. Cold weather varieties typically do not perform well in Texas because they do not retain sufficient acid content to ripen with the proper varietal character in the hot climate. Other varieties typically grown in more arid regions can be a nightmare of rot when grown in areas with summer rains. Make sure you investigate these characteristics before you plant.

In addition to varietal choices, there are often numerous clones within a variety that may be more or less adapted to a particular climate. Because there are such diverse growing regions in the state, a single variety trial experiment would offer insight into only the particular region where the experimental planting occurred. In addition, there are many, many new varieties making their way into the public sector from plant quarantine, and the number of hot weather varieties that will be available to nurseries in the near future will be bewildering. It will be tempting to be a "pioneer" and be the first to plant commercial acreage of a new variety. Remember what happens to pioneers, however; a few end up with the Ponderosa, but many more end up like the Donner Party.

Pierce's Disease Resistant and Tolerant Varieties

One of Andrew Walker's Pierce's disease resistant hybrids, U05–02–35.

As of 2014, there are only a couple of varieties available to growers seeking to harvest high-quality wine grapes in very high PD risk areas. In ten years, that will change drastically. The efforts of Andy Walker, grape breeder at the University of California, Davis, to incorporate PD resistance with high wine grape quality has produced a progression of new selections with 87 to 97 percent *V. vinifera* parentage that have wine quality as high as any existing *vinifera* variety. Those selections are currently under evaluation in several locations in Texas, and some will make their way into the public sector in the coming years. For now, however, the varieties described below are what are commercially available that have acceptable wine grape quality.

'Blanc DuBois'

'Blanc DuBois'

Released in 1987 by John Mortensen from the University of Florida, 'Blanc DuBois' is undoubtedly the highest quality PD tolerant wine grape currently under cultivation. It is the result of a complex process of hybridization that began in 1946 and ended in the cross of Fla D6–148 × 'Cardinal.' 'Blanc' makes a beautiful white wine, with muted Muscat notes and wine quality more akin to Sauvignon Blanc. It is resistant to many bunch rot organisms, seemingly immune to powdery mildew, but very susceptible to anthracnose. 'Blanc DuBois' is grown in many parts of the state that have PD pressure ranging from modest to severe.

'Black Spanish'

'Black Spanish,' aka 'Jacquez,' 'Lenoir'

This grape, whose exact parentage is unknown, has been grown in the southeastern United States at least since and perhaps well before the eighteenth century. Described as either of *V. aestivalis* or *V. bourquiniana* descent, this grape is very high in acids, tannin, and color, which limit the value of the wine in commercial markets. 'Black Spanish' is a resilient grape, also apparently immune to powdery mildew but susceptible to downy mildew and bunch rot organisms, especially when rainfall occurs near fruit maturity. While some wineries produce a still red with this variety, most commonly 'Black Spanish' is used for the production of port. 'Favorite,' thought to be a sibling of 'Black Spanish,' produces wines of similar quality.

'Victoria Red'

While 'Victoria Red' may have a place as a neutral white wine blender, the fresh market potential of this grape will have a profound impact on home and limited commercial grape production in areas with high to moderate Pierce's disease pressure. Released in 2010 by the University of Arkansas, Texas AgriLife, and Tarkington vineyards, this grape has a complex lineage and is the result of a 1971 cross: Ark 1123 × 'Exotic.' It has an attractive pinkish to red color, has large, loose clusters, and produces very large berries (eight grams on average) without the addition of any growth regulators, girdling, or cluster manipulation. It appears to be quite resistant to powdery mildew, and the loose cluster architecture makes it resistant to bunch rot organisms.

'Victoria Red'

Bordeaux Varieties

Texas is most certainly not the analog of the wine region of Bordeaux, France. Ripening conditions in Texas are much hotter, but because of varietal name recognition and, in some cases, very high wine quality, these varieties will continue to be important in the Texas wine marketplace. Color development in red Bordeaux varieties is much easier to achieve in the more northern parts of the state, but high-quality wine is also being produced in the Hill Country.

'Chardonnay'

'Sémillon'

'Chardonnay'

A versatile wine grape that is a standard Bordeaux variety (many consider this grape to be from Burgundy), 'Chardonnay' has yielded very high quality wine in many parts of the state. The greatest challenge to growing 'Chardonnay' is its tendency to break dormancy very early and be subject to crop loss due to spring frost. If grown with care, this variety has sufficient cold hardiness for the Texas climate but must be planted on sites with superior air drainage.

'Sémillon'

'Sémillon' is the major white grape variety grown in Bordeaux and is known for keeping high fruit quality while bearing crops of six tons or more per acre. Rather low in acidity, 'Sémillon' is commonly blended with other white varieties to produce high-quality, complex wines. While wine quality can be excellent from this grape, it is a challenge to grow due to the compactness of the cluster and its propensity to rot from split berries or rainfall near harvest.

'Merlot'

When cropped at levels of approximately four tons per acre, exceptional Merlot wines have been made in many areas of the state. Above those cropping levels, 'Merlot' grapes fail to achieve the adequate color and flavor intensities needed for premium wine markets. Excessive moisture near harvest can also greatly reduce the ability of this variety to produce the color needed for still red wines. Because of its broad consumer appeal, 'Merlot' will continue to be grown in most growing regions of the state.

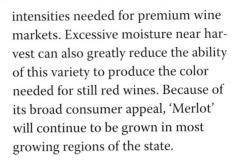
'Cabernet Sauvignon'

'Cabernet Sauvignon'

The red wine with the broadest consumer appeal, 'Cabernet Sauvignon,' or 'Cab,' is a little tougher to grow into a premium wine in the warmer parts of the state. Excellent wines have been made, no doubt, but with excessive crop loads or less than ideal ripening conditions, 'Cab' will not produce adequate color for a varietal wine. Some believe that the best 'Cabernet Sauvignon' grapes are grown in the High Plains, under cooler ripening conditions, while others think that the best Cab wines are made from High Plains fruit blended with fruit from other growing regions.

'Malbec'

A finicky grape to grow and produce consistently, the Texas 'Malbec' variety has been made into superior wines, either varietals or blends, from fruit grown in many parts of the state. With good acclimation, 'Malbec' is plenty cold hardy, but it de-acclimates quickly and can be sensitive to spring frosts and freezes. Many believe that the superior wines made from this variety warrant the additional risk, however.

'Cabernet Franc'

'Cabernet Franc'

The most cold hardy of the Bordeaux varieties, 'Cabernet Franc,' or 'Cab Franc,' is probably best grown only in the High Plains. While useful as a blender in other areas, this grape simply refuses to produce acceptable color unless grown under cool ripening conditions.

Rhône Varieties

The Rhône Valley, in southern France, is home to varieties that have gained popularity among Texas grape growers and wine makers. These hot weather varieties tend to ripen with much greater varietal character in the Texas climate as compared to varieties more adapted to cooler regions. Consumer name recognition is slowly gaining. Since approximately 2003, these varieties have been the source of some of the best Texas wines, many of which were blends from two or more of these varieties.

'Syrah'

'Syrah' ('Shiraz')

The different names for the 'Syrah' grape and its wines tend to reflect the wine style of the region. Syrah, the French wine, tends to be made in a more earthy, dry style, while the Australian Shiraz is produced in a much more fruit-forward fashion. While this variety can be a bit cold tender and finicky to set crops in younger vineyards, it has been and will remain a grape planted in much of Texas. 'Syrah' colors well, retains sufficient acids for a varietal wine, and is the backbone of GSM blends ('Grenache', 'Syrah', and 'Mourvèdre'). At present, some consider this variety to be slightly over-planted for current Texas markets.

'Grenache'

While it tends to retain beautiful acidity, 'Grenache' has a tendency to produce poor color in much of Texas. As a result, it has limited appeal as a varietal wine but is indispensable as a blender in GSM wines.

'Mourvèdre' ('Mataro')

An extremely upright, almost self-training vine, 'Mourvèdre' is a favorite among many growers. There have been some amazing varietal wines produced from this grape, but with nominal name recognition, it is most commonly blended with the other two varieties listed above. Like 'Syrah', it produces consistent color and retains acidity even in the hottest summers.

'Viognier'

The premier white wine grape of the Rhône region, 'Viognier' has made a place for itself in Texas because it is a variety that can consistently produce exceptionally high quality wines. This grape is a little tender and prone to wind damage but can produce good to high yields and still retain high fruit quality. The spicy wine produced from this grape is considered to be a hot weather equivalent of Gewürztraminer.

'Viognier'

Other Hot Weather Varieties

Other traditional hot weather grape-growing regions around the world have spent centuries weeding out varieties that do not consistently perform well relative to those that are well adapted. That said, just because certain varieties perform well in hot weather does not mean they will be great in the Texas climate, but it may mean they have a good chance of being more adapted than not. Grower experimentation with alternative grape varieties has led to small block establishment in some vineyards. Through this process the Texas wine industry has diversified into wider cultivation of lesser known grape varieties. To some, these lesser known varieties may be considered the standard of a particular winery, while others consider them a gamble. Consumer education will be critical for widespread acceptance of these lesser known varieties.

'Tempranillo'

'Tempranillo'

Grown on the Iberian peninsula for centuries, 'Tempranillo' is the primary wine grape of the Rioja region of Spain. Outstanding wines from this hot weather variety have made it a mainstay of many growers and wineries in the state. Deep, dark color, acid retention, and soft tannin structure have made these wines approachable and appreciated by the consuming public. 'Tempranillo' has found a home in Texas.

'Sangiovese'

Like other hot weather Italian varieties, 'Sangiovese' is well adapted to the Texas climate and has produced multiple double gold medal wines from Texas wineries. In Texas, 'Sangiovese' produces wines with a distinct garnet color, tart fruit notes, and very mild tannins. It is also commonly blended with other red varieties to produce more full-bodied wines. The down sides to growing this variety are a tendency to break dormancy early (like 'Chardonnay'), relative cold sensitivity, and high susceptibility to downy mildew.

'Sangiovese'

'Vermentino'

Widely grown in Sardinia and to some extent Corsica, 'Vermentino' is a hot weather Italian variety that has produced some outstanding wines in Texas. A late ripening variety, it produces large, loose clusters and is somewhat resistant to rot, allowing the fruit to hang to full maturity. This variety is not especially hard to grow and may well become a standard white variety for Texas.

'Petit Mansang'

While not widely planted in Texas at this time, 'Petit Mansang' is widely grown in Virginia, where it is considered to be a standard blender for many white wines. With high acid retention and loose clusters, this grape from southwestern France deserves some attention in future plantings.

'Vermentino'

There are numerous other varieties that have produced high-quality wines, and they include additional Italian varieties such as 'Nebbiolo' and 'Nero d'Avola'; Portuguese varieties 'Touriga Nacional', 'Alvarinho', and 'Tinto Cão'; the French 'Petit Verdot' and 'Tannat', and the list goes on. The Texas wine industry has made giant strides in varietal selection and winemaking since the early 1980s. The next thirty will undoubtedly be years of exciting growth.

Choosing a Training System

A TRAINING system is a systematized approach to arranging both permanent and annual parts of the vine to optimally intercept light and produce a crop. In nature, grapevines climb into the canopy of trees and other structures where they search for light and naturally reach a balance of growth and fruit production. Commercial grape growing utilizes specific training techniques that optimize sunlight interception, allow for manipulation of the canopy and application of protective materials to optimize fruit yield and quality, and allow for efficient harvest of the crop. Throughout the world, different training systems have been developed to fit local conditions of climate, available labor, and the specific varieties grown in that region. Although an exotic training system may seem appealing, prospective growers should carefully weigh the benefits and potential liabilities inherent in those systems before adopting a novel technique.

Cordon Training

Most *vinifera* wine grapes have erect growth habits and are generally trained to a mid-wire training system. In such a system, the trunk is trained to a cordon wire about thirty-six to forty-two inches above the vineyard floor. From there, cordons, or lateral, semipermanent extensions of the trunk, are established in one or both directions down the wire. The word *semipermanent* is operative here because, contrary to common thinking, cordons need to be renewed on a regular basis. From these cordons, a single vertical canopy is developed for sunlight interception and fruit production, and these undivided canopies are commonly called single curtain systems.

Cordons can be developed from a trunk and extend in both directions down the row (a bilateral cordon), or, if vines are spaced closely together, they can all be trained with a single cordon in the same direc-

Here, a vigorously growing shoot has been laid down to form a cordon and has been tipped to force lateral growth for the next year's fruiting spurs.

A young dormant vine with bilateral cordon established and sufficient cane growth to develop fruiting spurs.

An older unilateral cordon block that has been rough pruned in early spring.

tion down the row (a unilateral cordon). In each case, spurs (also known as outlets), are established to spatially separate the year-old wood responsible for fruiting each year. Like cordons, spurs are semipermanent parts of the vine that need renewal every few seasons. The shoots that arise from these spurs are trained vertically upward, where they bear fruit and develop the canopy necessary for optimal photosynthesis. Cordons are replaced or renewed because mechanical damage, winter injury, fungal canker infection, or girdling from the cordon wire or even tendrils can cause a lack of vigor in a cordon, resulting in a loss of productivity. In most cases, when renewing cordons, healthy canes can be selected from close to the head region of a vine to replace the old unproductive cordon, and a higher number of nodes can be left on the remaining cordon after pruning so that vine yield is not excessively reduced in any one season.

For many American or French-American hybrid varieties, shoot growth is naturally downward (procumbent), so cordons are typically established five to six feet above the ground (a high-wire cordon) to provide enough space for an adequate canopy to develop downward toward the ground. The principles behind both high-wire and mid-wire cordon systems are the same: grow healthy trunks and cordons and develop and maintain healthy canopies. The difference between the two types of systems is that the canopies are developed in the orientation the varieties naturally grow.

Cane Training

Some growers choose to employ cane training systems rather than cordon systems. In a cane pruned system, a trunk and head region are developed on each vine, but rather than developing cordons, the vine is pruned back to a designated number of canes for annual fruit production. Thus, if growers expect to leave, for example, twenty high-quality nodes after winter pruning, then, rather than distribute those over ten spurs with two nodes each, they will choose to leave two canes with ten nodes each. Canes are then tied to the trellis, at times directly to the training wire or at times bent over a higher wire, then tied down to the training wire. This bending of canes can assist in uniform budbreak, which translates into uniform fruit maturity.

One advantage of cane pruning is that, without cordons, the risk of infection from fungal canker pathogens, such as eutypa, bot canker, and phomopsis cane and leaf spot, decreases. The downside of cane pruning is that it is harder to train a labor force to select high-quality canes than it is to manage a cordon system. High-quality canes appear bright in color, are typically between one-quarter and one-half inch in diameter, and have short internode length. Canes that are dull in appearance and have wide internode distances were typically shaded during the previous growing season and have reduced hardiness and fruitfulness. Canes exceeding seven-eighths of an inch in diameter are considered "bull canes." They are usually relatively unfruitful and have irregular budbreak. Cane pruned vines must be not only pruned but tied on a timely basis (before bud swell) to avoid breaking primary bearing shoots. Although it is more labor intensive to cane prune vines, some growers believe that specific varieties perform better under this system.

Divided Canopies

In some locations, and especially with very vigorous vines, a single canopy does not adequately intercept sunlight and take advantage of the yield potential these large vines present. In the mid-1960s, Nelson Shaulis introduced the Geneva Double Curtain (GDC) training system, and he described it as "an aerial simulation of narrowly spaced rows." The system was designed specifically for the *labrusca* varieties such as 'Concord' and 'Niagara' but has been adapted to many other procumbent varieties, including 'Norton' and 'Catawba.'

The limitation of spacing rows closer together was that there was

◄ In addition to a wire for the irrigation pipe and a cordon wire, varieties trained with vertical shoot positioning typically have two or more pairs of catch wires that allow shoots to grow erect and still allow sufficient sunlight exposure to the fruiting zone.

▼ A high-wire cordon system used for some American hybrid varieties.

Older cane pruned vines immediately after pruning. Grapevines are the most heavily pruned of all perennial fruiting crop plants.

Multitrunked *vinifera* block that has been cane pruned and tied back to the cordon wire.

no equipment at that time that could navigate four-and-a-half-foot row spacing. By staying with the traditional spacing of nine feet by eight feet, every other vine would be trained to alternate sides of a two-wire cordon system. However, in that traditional arrangement with the vine alternating between two training wires, each vine can develop a cordon eight feet in each direction, allowing for a greater space to train large vines. To achieve the goal of simulating two independent canopies, GDC training requires that shoots growing between the two canopies be positioned so that there is distinct separation and full light penetration between the two canopies. Excessive downward positioning of vegetative shoots can lead over time to a reduction of vine size and yield.

Similar divided canopy systems, such as goblet and open lyre training systems, have been established for erect *vinifera* varieties. The concepts

Dormant Geneva Double Curtain block of grapevines. This system is used for American or hybrid grapes with a procumbent growth pattern.

An open lyre divided canopy training system for erect-growing *vinifera* varieties.

are the same except that the alternating cordon wires are approximately forty inches above the vineyard floor and, as most *vinifera* are apt to grow, the annual canopy is trained upward. Like GDC, this system is efficient at capturing sunlight and increasing yields and quality only if the canopies are kept separate through shoot positioning and removal. While divided canopy systems can be mechanically harvested, most lyre system plantings are still harvested by hand.

Pruning and Training Dormant Vines

RAPEVINES are generally pruned more than any other perennial fruit-producing plant. It's important for a grape grower not only to know how to prune vines but also to fully understand why it must be done. In all but minimally pruned training systems, annual dormant pruning is needed to invigorate growth the following spring, regulate crop load, and manage light interception by the fruit and by leaves that are in proximity to the ripening fruit. Dormant pruning of all temperate woody plants is an *invigorating action.* The more severely plants are pruned, the greater the compensatory growth that plant will exhibit the following growing season.

Pruning grapevines can be a daunting task at first. They are the most heavily pruned of all perennial crops.

With grapevines, the goal is to prune vines so that the remaining number of fruitful nodes on spurs or canes is sufficient to grow an adequate, healthy canopy with an appropriate crop for the age and size of a vine. If too few fruitful nodes are left, then the vine may exhibit excessive vegetative growth and produce a less than desirable crop. The excessive vegetative growth could result in excessive shading of the renewal zone, causing poor bud fruitfulness the following year, which encourages even more vegetative growth and even more shading. In addition, the shading will probably result in a crop with low color or soluble solid accumulation. If too many nodes are retained, an excessive crop will be set, resulting in low vegetative vigor and poor canopy fill unless some of the crop is removed in the month following bloom. In this case, the lack of sufficient, healthy canopy will result in a crop that will not ripen properly and a grapevine that has been severely stressed.

Vines that have been overcropped are exceptionally susceptible to winter injury, as well as infection from fungal and bacterial pathogens. Stressed vines also typically have reduced bud fruitfulness, which reduces the cropping potential for the following year. As in much of viticulture, properly pruning a vineyard represents the fine balance that is both the art and science of producing high-quality crops in healthy, long-lived vineyards.

Principles of Balanced Pruning

The three principles of balanced pruning are node (bud) number, bud quality, and node (bud) distribution. There is no standard number of nodes or buds to leave on a grapevine. The proper node number to leave after dormant pruning will vary by grape variety, vine age, vine size, and environmental and site influences, as well as the fruit quality desired by the buyer of the crop. Growers must learn how each variety performs on their vineyard site and how to adjust annual pruning to compensate for environmental influences.

Vine size directly reflects the bearing capacity of individual vines on a year-to-year basis. Nelson Shaulis developed the concept of balance pruning to help grape growers predict the bearing capacity of individual grapevines and to properly prune them, both to produce a healthy canopy and to ripen the crop. Growers begin this process by rough pruning a grapevine and weighing the removed year-old dormant wood. If older trunk or cordon wood is pruned out during this process, that weight is not counted; only growth that was produced the previous growing season is included in the dormant pruning weight. In rough pruning, typically 50 percent more nodes will be left than will actually remain after a vine has gone through final pruning, but it allows a grower to quantify the rough "pruning weight" of a vine or block of vines. Healthy, mature vines typically have a pruning weight of between one-quarter to one-half pound per linear foot of row (although with divided canopies this figure may be higher).

Vines with lower pruning weights could have been overcropped or have suffered from insufficient water, fertilizer, or weed control. Vines with excessive pruning weights could have been undercropped, had excessive nitrogen or irrigation, or gone through a season of exceptionally high rainfall. Vines with low pruning weights are not capable of producing a reasonable crop of quality fruit, while vines with excessive pruning weights are typically not especially fruitful because of shading of basal shoots. Vines with either very high or very low pruning weights are more vulnerable to winter injury.

Nobody expects growers to go out and measure the pruning weights of each vine every year, but it is important for growers to become familiar with pruning weights in order to gauge the cropping level and viticultural inputs of a vineyard block. Unfortunately, it is common for growers to continue to leave the same number of buds after pruning year after year in order to match the crop produced in a previous season even though over time vine size declines. When that happens, the vines are saying something: you are running them at a higher capacity than they are capable of sustaining, and something needs to change. Perhaps in a single year, crop reduction and increased nitrogen, water, and weed control can help a block regain vine size. Failure to recognize and react to this phenomenon commonly results in winter injury or vines dying or declining from pathogens that may not otherwise be lethal.

Bud quality is an important consideration when pruning grapevines, and there are several visual cues that a grower can use to reserve fruitful buds and canes during dormant pruning. Distance between nodes (internode length) is one such indicator, and an internode length of three inches is considered ideal. In order for a bud to be fully fruitful, it should have been grown on a cane that was properly exposed to sunlight. Cane color is another indicator. Retained dormant canes and spurs should be bright in color, not dull, and should be either tan or mahogany in color. Pale, dull canes and spur positions are indicators that those places were shaded during the past season. The third indicator is cane size. Canes and spurs should ideally measure between one-quarter to one-half inch in diameter. Smaller diameter canes are commonly only modestly fruitful, while large canes are typically very vegetative in nature and are perhaps not fruitful at all. These large canes, which can exceed ten feet in length, are referred to as "bull canes" and will not yield canes that grow fruit. Bull canes also won't yield canes appropriate for generating new cordons. When growers allow for

Avoid retaining excessively vigorous "bull" canes for either fruiting spurs or cordons. The "bull" wood is typically cold tender and unfruitful. The smaller diameter wood (*in hand*) would be more appropriate for either spur or cordon.

proper bud distribution it means that buds retained after winter pruning are evenly distributed in the space allotted for that vine in the trellis. This practice ensures that shoots arising from those buds have the greatest chance of receiving full sunlight and not being shaded. Good bud distribution is important for the production of the crop that season, as well as for maturation of the canes that will grow a crop the following season. Occasionally, though, this practice can be altered. For example, if a cordon is being replaced and a new cane is tied down where an old cordon was cut out, a greater than normal bud number can be left on the other side of the vine to make up for the shortfall on the renewed side.

Delayed Pruning

In addition to knowing how to prune vines and why it is necessary, growers must address the question of when to prune. For dormant pruning, the answer is as late as possible because pruning is one of the strategies to reduce the probability of economic loss due to spring frost or to adjust the number of retained nodes to compensate for winter injury. Delaying pruning is much more important for mature vineyards, from which growers expect a crop, than for young, nonbearing vineyards.

When grapevines break dormancy, they normally start growing at the apical end of dormant canes, with budbreak slowly progressing back

This mature dormant vine had spurs develop uniformly across its space in the trellis, making pruning and bud distribution a much easier task.

toward the basal buds of a cane near the cordon. If we prune early, the growth begins on the apical buds, which become the buds we want to retain for the following season. Delayed pruning, or double pruning, is a technique by which very rough pruning is done with the intent to come back later in the spring and adjust the bud number and distribution to what is actually desired. In double pruning, spurs are left seven to eight nodes in length, undesirable growth is removed, and canes are tied down for cordon renewal or extension as needed.

Double pruning allows for less-skilled labor to come in and pull brush from the trellis and repair or tighten the trellis as needed. Growers are then in a position to come back when the risk of spring frost is much less, conduct final pruning and bud number adjustments, and adjust and force vine growth at a time when losses from spring frost are far less likely. A strong word of caution is in order, however: the period a grower can wait to come back and adjust is limited. If rough pruning has not been adjusted to final spur length and growth is initiated from apical buds, at some point the fruitfulness of the basal buds you intend to leave is lost. Certainly growers can wait for growth on undesirable wood to reach four to six inches in length before coming back for final adjustment, but once growth exceeds that stage, the consequences cannot be predicted. Even the

This unilateral cordon vine had been rough pruned so that brush could be pulled and trellis wires tightened. As the weather warms in the spring, terminal buds will swell and begin to grow and the growth of desirable basal buds will be delayed. After the risk of spring frost has (presumably) passed, final pruning will be done to the desired length of spur. Delayed pruning is the greatest tool a grower has in reducing the risk of crop loss due to spring frost.

best grapevine anatomists cannot say at what point the floral primordia are lost and the buds are no longer fruitful. While the double-pruning technique is widely practiced and extremely valuable in reducing the impact of frost, it has inherent risks.

There are a multitude of training systems employed around the world to meet specific varietal, cultural, and environmental conditions. In Texas vineyards, unless vines are planted at a very high density, the most common method of training grapevines is in a bilateral cordon system. In this system, one or two trunks are trained to the cordon wire, and two mature canes, one traveling in each direction from the trunk, are tied down to the training wire to be retained as cordons. From these cordons, shoots are trained vertically to bear fruit and form the canopy. In the second year, these shoots, which are now canes, are pruned back to two or three nodes and are called spurs. Buds on these spurs become shoots that bear fruit annually. Spurs should be evenly spaced along the cordon in order to facilitate optimal light interception and full use of the trellis. If vines are closely spaced, some growers choose to train vines to a unilateral cordon system, in which arms are trained only in a single direction from the trunk.

Basic Pruning Techniques

The following suggestions for pruning and training young vines are intended for growers of *vinifera* and hybrid varieties with strong vertical growth. American hybrids tend to have a procumbent growth pattern (growing downward rather than upward) and are trained in a similar way, but to different training systems. Variations on vine type and alternative trellis configurations will also be addressed.

Year One

No matter how old the nursery stock is at planting, a young vineyard in the first year of growth after planting is considered to be in "first leaf" and requires special care. Because grapevines lose a portion of their root system when they are dug up at the nursery, the root system is thus somewhat compromised. However, cutting back the vine to two or three buds above the graft union after transplanting will mean that the vine and root system will not have to support as much canopy. The vine will then have a chance to recover and once again reach equilibrium between root growth and canopy.

Vines will grow quickly in the spring and should be trained up a stake or twine tied to the trellis. "Grow tubes" can also be used. Several types of grow tubes are on the market, and they range from polyurethane or polymer translucent tubes to opaque, paraffin-coated cardboard (similar to the material used in milk cartons). New growers will get widely varying advice on how to train young vines, so it's important to understand the principles of plant growth and act accordingly.

While the practice in many vineyards is to train a number of grapevines to a single trunk, retaining two trunks (double trunking) has many advantages. If a cold event causes freeze injury or if a single trunk is infected with a fungal or bacterial pathogen, a grower can remove the damaged trunk and retain the healthy one for fruiting while a replacement trunk is grown for subsequent seasons. Most growers are content to grow one or two healthy shoots the first growing season, and these shoots will become the trunk system.

Growers will voice a variety of opinions regarding how to deal with shoots from the vine other than those that will be left to serve as trunks. Some advocate cutting off these "extra" shoots, arguing that the vine should put all energy into the trunks that will be retained. Others insist on severely pruning a strong vegetative shoot to encourage lat-

▲ A mature vine trained with two trunks. This technique gives growers more options when faced with freeze injury or wood-rotting canker infection.

▶ A vine after the first growing season.

After one year of growth, a new vine can be pruned back to one or two trunks pruned at the cordon wire.

eral growth for cordon development. As was previously stated, dormant pruning is an invigorating action, but the converse rule is that pruning off photosynthetic tissue during the growing season is a dwarfing action. A vine expends energy to grow a shoot, and if that shoot is removed before it can replace that energy through photosynthesis, it is a net loss for the vine. It is thus strongly recommended that growers remove only green tissue that is problematic for the developing vine. For example, if shoots arise from the rootstock, they should of course be removed. Shoots being retained for new trunks should take precedence and be trained up the stake or string; they *must not be shaded* by other growing shoots because they are there to intercept sunlight and provide energy to the young vine. If shoots are growing on the ground and making weed control problematic they should removed, but the greater the photosynthetic capacity of a first-leaf vine, the more it will grow in response to dormant pruning the following year.

▶ This dormant vine had a vigorous cane cut back severely during the growing season in an attempt to force lateral growth for cordon development. Removing large amounts of photosynthetic area during the growing season severely dwarfs growth that season and results in a vine that is difficult to train.

▼ Following first dormant pruning, new shoots force and can be developed into cordons.

Year Two

133

Pruning
and Training
Dormant
Vines

Shoots (one or two) grown for trunks should be pruned back at cordon wire height or, if vigorous enough, tied to the cordon wire to begin the formation of cordons. All other shoots should be pruned back flush with the trunk. As spring approaches and vines begin to grow, rub off all newly emerging shoots that arise from ground level to six inches below the cordon wire by the time they are one inch long. This practice will allow the higher shoots to grow into cordons. The grower should take two shoots that arise from the trunk(s) near the cordon wire and train them in opposite directions down the wire. Some tying may be needed to keep these shoots on the wire, but tendril growth will most probably do that job. In the Texas climate, these shoots exhibit tremendous apical dominance and grow as much as twenty feet in a given season if not trained differently.

In order to force lateral shoots to form spurs for fruiting in the third leaf (or third year of vine growth), new cordon shoots should have three to four inches of the shoot tip removed when they reach about fifteen inches in length. Typically the most distal bud will start growing immediately and regain apical dominance, but several of the other lateral buds will also start to grow. This process is repeated again when the new cordons reach two feet in length and again when they reach three feet in length. Remember, the idea is to avoid compromising the photosynthetic capacity of the canopy of a young vine but to make enough small adjustments to shoot growth to overcome the apical dominance a vine would otherwise exhibit.

Under certain conditions, growers can push growth of young vines to the year two stage in the first growing season. At times, growers have achieved a crop from second-leaf vines while getting a good balance between vine growth and productivity. This technique should be attempted only with superior vine growth and vineyard management, and cropping levels should be kept quite modest to avoid compromising the growth or winter hardiness of young vines.

Year Three

A young vineyard is typically ready to bear its first crop in the third year. During dormancy before the third growing season, prune off all canes on the one or two permanent trunks that arise below the cordon wire. Canes to be retained for new spurs should be four to five inches apart in each direction along the cordon where they exist. Most viticulturists look for three to five bearing shoots per foot of canopy.

In divided canopy systems there are two linear feet of canopy per foot of vineyard row. Weak or excessively vigorous canes should be removed. Canes one-quarter to one-half inch in diameter with bright color and close internode spacing make the most fruitful spurs.

Spurs should be upright or nearly upright to promote even canopy growth and fruiting. Spurs should be cut back to two or three nodes. The proper number of nodes left after dormant pruning is dependent on the desired crop load. Typically, each fruitful node will produce a shoot with two clusters, but different varieties produce clusters that vary considerably in weight. For example, 'Merlot' and 'Cabernet Sauvignon' typically produce clusters that weigh approximately one-third of a pound, while 'Sangiovese' and 'Trebbiano Toscano' can produce clusters weighing in excess of one pound. With cluster weight in mind, determine whether the variety planted should have fewer nodes (for heavier clusters) or can handle more nodes (for lighter clusters).

Mature Vines

Once a vine has been established with a strong trunk, cordons, and spurs, annual pruning becomes a little more routine, with selection and renewal of spurs to leave the appropriate number of nodes for a full crop. When choosing year-old wood to retain for fruiting, keep in mind that the goal is to keep the fruiting or renewal zone as close to the cordon as possible Choose the most basal, highest quality cane for retention as a spur to keep the renewal zone from creeping away from the cordon.

Although we normally think of trunks and cordons as permanent parts of the vine, over time, they too may need to be renewed or replaced. Winter injury, mechanical injury, or fungal canker organisms may compromise the vascular system of trunks and cordons, requiring growers to replace them with a new one-year-old cane. Avoid the temptation of waiting until a vine is in severe decline; instead, consider this renewal process to be a normal part of the annual pruning cycle. If fungal canker infection has progressed into the trunk, renewal efforts may not yield the desired results.

Other Training Systems

There are numerous training system alternatives used around the world for different reasons. Carefully consider the consequences of these other systems because there could well be varietal or climatic limitations to their use in the Texas climate. Many Pierce's disease tolerant varieties

A third-leaf vine before dormant pruning.

A third-leaf vine after dormant pruning.

A dormant mature vine that fully developed prior to winter pruning.

Developing and maintaining spur positions is a challenge for new and experienced growers alike. Choose more basal dormant canes for fruiting wood to keep the "renewal" zone as low to the cordon as possible.

A well-developed spur just prior to budbreak.

Suckers that arise above the graft union can be maintained to replace old or winter injured trunks.

such as 'Blanc DuBois' and 'Black Spanish' are commonly grown on a high-wire cordon with spurs and shoots positioned downward. This system is employed because of the natural downward, or procumbent, pattern of growth.

Before adopting a novel pruning or training system, new growers should carefully weigh the reasons such a system is not currently being employed in their region. Growers who engage in cane pruning, which requires tying the canes to the trellis, must be aware of complicating matters. Not only is it more challenging to train vineyard workers to select optimal canes than it is to spur prune vines, but, with the common practice in Texas of delaying pruning until at least bud swell, many buds may be knocked off of the cane during the tying process.

For very vigorous vines, usually those with annual pruning weights in excess of five pounds, divided canopies can be used to manage growth and optimize light interception. The Geneva Double Curtain system was developed for high-wire training, and the open lyre and other systems are

Pruning is the marriage of the art and the science of viticulture. Working with experienced viticulturists is the best way to learn the basics.

used with varieties trained on a mid-wire trellis. Divided canopy systems, described as aerial simulations of narrowly spaced rows, allow growers to employ the same equipment used for single canopies but to optimize light interception as would be the case with more narrowly spaced rows. Any divided canopy system is dependent on positioning shoots during the growing season in order to prevent shading. While divided canopy systems can improve both yield and fruit quality, they are more labor intensive to maintain.

Canopy Management

THE TERM "canopy management" refers to a number of cultural practices that influence how the vine grows and how the fruit ripens during the growing season. The goals of canopy management can be succinctly described as improving fruit quality and maintaining grapevine health and productivity. These goals are achieved by managing vegetative vine vigor, optimizing sunlight exposure to the fruiting and bud renewal zone, and manipulating the microclimate within the shoots.

As with any perennial crop, vineyard canopy management techniques are only a subset of the cultural management practices necessary to strike the fine balance between vegetative growth and high-quality fruit production. On a mature vineyard, it is common practice early in the spring to remove all shoots arising from ground level to six inches below the cordon wire. This pruning is best accomplished when shoots are one to two inches long, but some growers wait until there is growth of six inches or more before removing these shoots. However, if trunks or cordons need to be replaced, the suckers or sprouts above the graft union that commonly arise in the spring can provide the replacement. Simply retaining one strong sucker and training it alongside the existing trunk is common practice. If there is a profusion of suckers arising low on the trunk, it could well be that some winter injury has occurred higher up on the trunk. In that case, retaining two shoots, at least through the early summer, may be wise should cordon replacement be needed for the following year.

Shoot Thinning

There is a general consensus among growers that the ideal shoot density on a grapevine is between three and five shoots per linear foot of row.

▲ On mature vines, shoots originating below the cordon wire are removed early in the season in the first act of managing the canopy for the growing season.

▶ Prolific sucker development is a sign that there is a "plumbing problem" above that point. Typically, freeze injury forces growth from uninjured portions of the plant. One or two strong shoots should be retained and trained for trunk and cordon replacement.

This limitation not only prevents overcropping the vine but also ensures that the remaining shoots have full sun exposure and are not shaded. This full sun exposure provides the ideal light environment for fruit ripening, high photosynthetic efficiency in the basal leaves of shoots, and high fruitfulness for the buds that will ultimately become the fruiting wood for the following season.

For vines spaced six feet apart on an undivided canopy, growers should strive to produce eighteen to thirty primary bearing shoots per vine. On vines spaced at eight feet, a total of twenty-four to forty shoots per vine is most favorable. For productive, large-fruited varieties with large clusters, such as 'Sangiovese,' the lower shoot number is more desirable, while for varieties with smaller clusters, such as 'Pinot Gris,' the higher number is more appropriate. During annual dormant pruning, growers often leave more primary buds per vine than they ultimately want, in case a portion are lost to frost or hail. In addition, some vines produce shoots from latent buds in the cordon, giving rise to green shoots with no productivity. Where the canopy is sparse and the shoot number is below the optimal level, some of these shoots are commonly left to fill in the gaps and provide fruiting wood for the following year. When the density of either fruitful or unfruitful shoots exceeds what is desirable, shoot removal is conducted at a point when the grower feels comfortable that further shoot loss from natural elements such as wind or hail is unlikely. Shoots are tender and easily removed without shears when they are from five to ten inches in length, and it is in this time frame when most growers normally thin shoots. After this point in the growing season, shoots are more lignified, and removing them will require hand shears and considerably more time to accomplish.

This is the dilemma growers face each year: the earlier the shoot thinning takes place, the better the results but the greater the opportunity for shoots to be lost due to weather events. It is important to remember that removing green tissue from a vine is ultimately a dwarfing action, and the larger the shoot, the greater the dwarfing effect. If vines are excessively vegetative, shoot thinning can be a useful tool for bringing vine vigor back into balance. If vines are struggling, however, late shoot removal can result in vines that are both incapable of ripening a high-quality crop and susceptible to pathogens and cold injury. Shoot thinning is an exercise that is not necessary in every year, nor appropriate in every vineyard, but it is a practice that can help maintain vine balance.

Shoot Positioning

When shoots are young, they are very tender and easily broken because lignification of tender tissue has not yet occurred. Lignification is defined as the process of green plant tissue becoming woody, and it occurs as a result of lignin deposition in plant tissue. In grapevines, this process becomes noticeable about the time that vines bloom. As a result, manipulation of the canopy before bloom brings with it the risk of shoot breakage. To avoid that problem, many growers use a catch wire or a pair of catch wires positioned about a foot above the cordon wire. With this practice, the tendrils of vertically growing shoots adhere themselves to or between the wires. Grapevines are natural climbers, and most shoots will naturally grow vertically along these wires to create a curtain of foliage. There are, of course, shoots that do not find their way to these wires and will bow away from the trellis. They are commonly in the way of equipment passage and can impede spray penetration into the fruiting zone.

It is common to have one or two more single or pairs of catch wires above the first to assist grapevines in growing a vertical hedge of foliage. Shoot positioning is nothing more than putting wayward shoots back where they belong, between the catch wires. Shoot positioning can begin at about fifteen to eighteen inches of shoot growth, but because shoots are still tender, only limited positioning can be achieved without undesirable shoot loss. After bloom and during fruit set, it is a common practice for these shoots to be manipulated and positioned between catch wires or tied to a series of catch wires. Achieving this moderately dense,

Shoot positioning is a laborious and time-consuming task.

A vigorously growing canopy in need of canopy management. If left unpositioned, shoots and fruit will be shaded, thus encouraging a microclimate conducive to fungal disease development.

vertical wall of foliage will efficiently intercept sunlight for efficient photosynthesis, allow for excellent airflow within the canopy to reduce fungal disease pressure, and expose both fruit and basal foliage to light needed for fruit quality and high bud fruitfulness.

In divided canopies, shoot positioning also means removing or moving shoots inappropriately growing upward or downward (depending on the training system) between the two canopies so that light may be fully intercepted by both canopies. Increased light interception is the key to making divided canopies work well and increasing yields and fruit quality. Division of canopies has traditionally been accomplished with manual labor, but because of the decreased availability and increased cost of a dependable labor pool, mechanical means of positioning grapevine shoots have been developed. While great strides have been made in the ability of these machines to position shoots with minimal vine impact, there will still be some breakage. Equipment design and, more importantly, operator skill will mitigate the negative effects of mechanical shoot positioning.

In the hot Texas climate, exposing fruit to sunlight on the eastern side of the canopy results in improved quality without the risk of fruit sunburn from western exposure.

With a shaded canopy, basal foliage senesces prematurely and adjacent buds can have reduced fruitfulness.

A mechanical aid for positioning and breaking open the center of GDC-trained blocks. Operator skill is the key to accomplishing the task with minimal vine injury.

Summer Hedging

When summer shoot growth passes through all of the catch wires and begins to topple back over the canopy, summer hedging must be employed to avoid canopy shading, which can have negative implications on the leeward side of the canopy. Hedging typically consists of trimming off the tender vegetative growth that extends six to twelve inches above the posts. In some areas, and with some varieties, growth of lateral shoots arising from the primary bearing shoots is common, so hedging can also include vertical cuts to reduce the density and thickness of the horizontal canopy. Under vigorous growing conditions, this hedging may be practiced three or more times during the summer, up to the point of harvest. Again, removal of green tissue will be a dwarfing action to the vines, but in some cases using this technique to slow vegetative growth may be warranted.

Timing is very important when it comes to hedging vineyard canopies during the summer. The goal of the hedging is to ensure that after véraison, and especially during the final phases of fruit maturity, the primary photosynthetic sink of the vines is the fruit, not the canopy. The carbohydrates produced during photosynthesis should be moving to the fruit for sugar accumulation, not to shoot tips for new growth. In more northern areas, the decreasing photoperiod in early autumn helps vines respond with a cessation of vegetative growth. In southern regions, where there

A *vinifera* block being summer hedged. To reduce the growth of lateral shoots, both top and side growth should be addressed.

is a relatively long photoperiod during fruit maturation, growers have a struggle. Especially in areas or seasons when there is tropical moisture falling, vines just want to grow. In arid areas, growers can withhold irrigation water to moderate vine vigor. In southern areas, although a drought may be in place, there may be tropical storms with heavy rains that result in untimely vegetative growth. There is no single recipe for accomplishing canopy hedging, so growers must see in their mind's eye what they want to achieve and do their best to accomplish that vision.

Practically speaking, there is a limited amount that growers can do to address this dilemma: summer hedging results in increased growth of lateral shoots, which in turn causes more shading. In growing seasons that produce very vigorous canopies, vertical hedging to remove these laterals may be necessary. Growers must understand the season, the variety they

are growing, and their site and, through trial and error, practice summer hedging to shift carbohydrate flow to the fruit. This principle is great
in theory but hard in practice. Understand the principles and promote
or curtail vine vigor whatever way you can. Create the best moderately
dense canopy and hedge when late-season growth occurs, but keep the
remaining canopy healthy. There is, again, no one formula. That is the
reason viticulture is both a science and an art.

Leaf Removal

Once fruit hit lag phase, approximately twenty-five to thirty days after
full bloom, many processes begin to take place within the berries and
shoots. Development of the primary compounds responsible for flavor
in berries is progressing, light exposure is accentuating color development in red-fruited varieties, and bud initiation that will generate the
next crop is winding down. Leaf removal is the plucking of foliage in the
lower seven to ten inches of the fruiting zone to reduce the leaf layer and
expose the fruit to more sunlight. Making sure that fruit and basal foliage
are well exposed to light optimizes all of these processes.

In most vineyards, leaf removal is accomplished with hand labor,
but there are mechanical leaf pullers that do a remarkable job. However,
when the fruit in a Texas vineyard reaches lag phase, the temperature
could be 100 degrees Fahrenheit or higher. Do we really want our fruit exposed to such extreme temperatures? No, not really, and that's why what
growers do in other areas of the country may not work in Texas. Dappled
light exposure will aid in color development and reduce the threat of late
season rot organisms, but fruit exposed to direct sunlight will absorb heat
and berry temperature will thus be several degrees higher than ambient
air temperature. Experience will dictate the extent to which leaf removal
will improve fruit quality on a given variety in a given location, but that
too will vary from year to year.

The concepts behind leaf removal are the same everywhere, but the
practice must be modified for the Texas environment. In southern regions where fruit ripens under sometimes brutal summer temperatures,
growers need to guard against excessive fruit exposure to the sun, which
will result in sunscald to the fruit. Thus, unlike growers elsewhere, Texas
growers should not remove foliage on both sides of the trellis; leaf removal on dense canopies is limited to the northern side of east-west
rows or the eastern side of north-south rows. This pattern of leaf removal
achieves the desired goals without the risk of fruit damage.

Leaf removal may mean reducing the density of the canopy to less than two leaf layers between the sun and the fruit, and, on protected exposures that do not face the heat of a Texas summer, it usually means removal of virtually all of the foliage in the lower seven to ten inches of the fruiting zone. This type of leaf removal seeks to accomplish the following goals:

- Improving fruit quality through the enhancement of color and flavor
- Increasing bud fruitfulness for the next crop by increasing the sun exposure of basal leaves that will be retained for the next crop
- Adjusting the microclimate of the canopy to reduce fungal disease pressure and enhance the ability of spray materials to adequately penetrate the canopy and protect ripening fruit

Not all vineyard blocks need leaf removal. When producers are growing rot-resistant varieties for the less-than-premium market, this practice is not cost effective. It also does not have as significant an impact on the quality of white varieties; the quality of red-fruited varieties is enhanced to a greater degree. Leaf removal is a tool, needed more during wet seasons than dry, but when growing tight-clustered varieties that are prone to *Botrytis* fungi or summer rots, it may indeed be a necessity. On others, it may be an option, in some cases, simply an unwarranted expense, and, in the worst case, could result in sunburned fruit that will yield low wine quality.

CHAPTER 16

Grapevine Nutrition and Vineyard Fertilization

T
O REMAIN healthy and productive, grapevines need to be supplied with the optimal amount of chemical nutrients over the course of each growing season. In order to properly fertilize a vineyard, growers need to understand how nutrients behave in the soil, how they move in plants, and how to adequately gauge the nutritional requirements of a specific block of grapevines.

The movement of nutrients in the soil depends on the chemical charge associated with an element or compound containing an essential element. Soil particles have chemical exchange sites where elements may be electrically bound or absorbed into the particle. These exchange sites are negatively charged, so ions with positive charges, such as potassium, magnesium, and phosphorus, are attracted and held by the soil. As a result, the soil holds these nutrients, and they resist movement due to leaching as a result of rainfall or excessive irrigation. Even among positively charged molecules (cations), elements are held more or less tightly to soil colloids and have differing rates of relative mobility in the soil. Nutrients such as nitrogen are typically found in the soil as negatively charged compounds (anions), which are not bound on soil exchange sites and consequently are subject to movement in the soil. Heavy rainfall can force mobile nutrients such as nitrogen out of the root zone of grapevines, and this fertilizing element is subsequently lost to leaching. Consequently, strategies for applying anion-type nutrients may be very different from those for cation-type nutrients. The seasonal needs of a grapevine for cation-type nutrients can be applied in a single application without fear of loss, while anion-type nutrients are commonly added in split or multiple applications.

The ability of a vine to absorb positively charged nutrients is dependent on:

- Nutrient availability, which may be limited by soil pH
- Adequate water in the soil so that nutrients may be extracted
- The presence and concentration of other nutrients that may result in competition for uptake

Nutrient Mobility in Grapevines

Just as essential plant nutrients differ in how they move in the soil, essential elements have different mobility patterns within grapevines. Some elements, such as nitrogen and magnesium, are highly mobile within vines. Thus, if the vine is in short supply of a mobile nutrient, it is capable of extracting it from older tissue and moving it to new shoot tips, where it is used to supply new growth. Because of this characteristic, deficiencies of mobile nutrients are first seen on older leaf tissue. Conversely, other nutrients, such as iron and zinc, are not mobile; the plant cannot take needed nutrients from older tissue and move them to support new growth. Deficiencies of immobile nutrients are thus seen first on new growth.

At times, growers try to correct nutritional deficiencies with foliar nutrient sprays. With immobile nutrients such as iron, the spray may "green up" some symptomatic tissue, but because iron is immobile, the next leaf that forms will be iron deficient. Correcting deficiencies of mobile nutrients like nitrogen with foliar sprays can at times be exactly the right solution because it keeps the canopy healthy without the burst of new vegetative growth a soil application may trigger.

Essential Plant Nutrients

Essential plant nutrients are categorized by the relative amounts of those nutrients that plants need. There are ten elements needed in large amounts. They are called macronutrients and include carbon, hydrogen, oxygen, phosphorus, potassium, nitrogen, sulfur, calcium, iron, and magnesium. There is still debate about the beneficial role of some elements needed in small amounts, known as micronutrients. Those commonly addressed in nutrient management plans are chlorine, boron, manganese, zinc, copper, and molybdenum.

Carbon, Hydrogen, and Oxygen

Carbon, hydrogen, and oxygen (C, H, and O) are essential to plant growth and development, but they are not elements we add to the vineyard; they are present in the atmosphere and the soil. Carbon, the building block of all organic compounds, is provided by atmospheric CO_2 taken into plants through stomata on the underside of leaves. Oxygen is a necessary metabolic component of all plants, either in its free form or as a by product of photosynthesis, when CO_2 and water are combined in the presence of light. Hydrogen, also a basic building block of all life, is provided by water, either in the atmosphere or, more commonly in plants, absorbed through the roots. Although these elements are common and abundant, they are mentioned here because they can be deficient in plants if soils are saturated with water or otherwise lack oxygen or if there are problems that limit photosynthesis. Growers do not need to worry about adding these elements, but choosing an appropriate site and following management practices that optimize photosynthesis ensure that plants have sufficient levels of these elements.

Nitrogen

Nitrogen (N) is an important component of plant proteins, amino acids, nucleic acids, and plant pigments. Nitrogen is highly mobile in plants, and, consequently, deficiencies first show up in older foliage. This element is also very mobile in soils, making it subject to losses under some environmental conditions.

More than any other element, nitrogen presents a management dilemma to new and seasoned grape growers alike. A grower must manage this element in order to effectively balance vegetative growth and fruit production. Because vines can be very responsive to nitrogen fertilization, it is all too easy to tip the scales in either direction. In almost every soil type, nitrogen applications are needed annually to encourage strong, vigorous growth, but soil types vary in their ability to hold nitrogen. In very coarse, sandy soils, where water moves freely, nitrogen may be best applied in many small doses.

Nitrogen fertilizers come in several forms but usually consist of nitrate, urea, and ammonium sources of nitrogen. Nitrate is the form of nitrogen most readily utilized by grapevines, but the nitrate ion is subject to several fates in the soil profile. Both urea and ammonium are positively charged ions, which theoretically can be held by negatively charged soil exchange sites, but both urea and ammonium are commonly broken down rapidly by soil microbes and are converted into negatively charged nitrate ions. With adequate water and oxygen in the root zone, new root tips readily absorb nitrogen. In either dry or waterlogged soils, grapevines often exhibit nutritional deficiencies that are a direct result of limiting soil conditions. When excessive moisture is present in the vineyard, nitrate ions are leached out of the soil profile when water drains. In saturated soils, nitrate ions can be chemically converted to nitrite, then nitrous oxide, and ultimately lost into the atmosphere as a gas through volatilization.

When supplied with adequate nitrogen and no other limiting factors are present, vines grow quickly in the spring to produce a lush canopy capable of ripening a full crop. If there is not enough nitrogen, vegetative growth is inadequate, internodes are very short, leaf color appears yellowish to pale green, and leaf photosynthetic capacity is reduced. This limited canopy will struggle to both ripen the crop and mature green shoots into overwintering canes.

While some growers depend on petiole sampling to determine if there are sufficient nitrogen inputs, others gauge their nitrogen levels based on canopy fill and leaf color. Applications of too much nitrogen result in an excessively vigorous canopy with long internodes, which shades developing fruit, creates a canopy susceptible to fungal pathogens, and results in poor bud fruitfulness for the following growing season. Excessively vigor-

Vines supplied with adequate nitrogen exhibit strong vegetative growth and have dark green foliage.

ous vines are also slow to harden off in the fall and are commonly more susceptible to winter injury than those with more appropriate nitrogen levels.

Nitrogen is generally recommended in units of actual nitrogen per acre. To convert this number to fertilizer rates, divide by the percentage of nitrogen in the fertilizer material. For example, an application of one hundred pounds of ammonium nitrate (33 percent N) would result in the application of thirty-three units of nitrogen; one hundred pounds of ammonium sulfate (21 percent N) will provide twenty-one units of nitrogen. Prilled urea (46 percent N) is used in some areas but is quite volatile at temperatures above 50 degrees Fahrenheit, and because it will be lost through volatilization if not immediately incorporated, it is not commonly used in southern vineyards.

In the past, many growers believed they had to apply nitrogen well before budbreak in order for vines to have adequate amounts of this nutrient when growth began. We now know that much of the nitrogen used early in the season is nitrogen that was absorbed the previous season and stored within the vine. This knowledge has allowed growers to change their fertilization practices in order to match canopy growth with crop load. Spring nitrogen applications are now commonly made after a grower

Nitrogen is very mobile in soils, and, in seasons with high rainfall, vines can become depleted of this vital nutrient. Nitrogen deficiency is first seen in the yellowing of basal foliage in the canopy.

feels the risk of spring frost has passed, and then supplemental applications are made during the course of the growing season as needed. Many growers now also make small applications in late fall in order for roots to have ample nitrogen to absorb over the winter.

Because in a mature vineyard roots functionally cover the entire vineyard floor, it is common practice to broadcast the first, and typically largest, application of nitrogen in the spring. Rainfall is needed to incorporate the fertilizer, and the chances of good rains are higher during the spring than later in the season. With reduced likelihood of rain later into the summer, supplemental applications of nitrogen are most efficiently applied by injecting through the drip system. Nitrogen additions are typically halted in July in order to discourage additional vegetative growth and allow vines to produce periderm and prepare for winter. If nitrogen levels appear to be insufficient to maintain canopy health going into the fall, foliar applications may help the vine regain leaf function without the risk of triggering a new flush of growth that a ground application might cause.

Potassium

From ancient times, materials such as manures and ashes have been added to crop land in order to sustain plant growth and productivity. In addition to nitrogen, potassium (K) is often a major component of manure and is present in fire residues as potash. Potassium is important in plants because it regulates protein synthesis, enzyme activation, ion absorption and transport, and solute transport and other aspects of vine water status. With insufficient potassium, vines struggle to achieve optimal soluble solids levels in fruit, and, under severe deficiency, fruit can drop from the vine prior to reaching maturity (a process known as shelling).

Potassium is very mobile in plants, and, consequently, deficiencies are seen first on older foliage and are characterized by basal leaves exhibiting interveinal chlorosis, with affected tissue progressing toward necrosis. On some red-fruited varieties, affected interveinal tissue may become red, bronze, or black. In 'Concord' vineyards, potassium deficiency is commonly referred to as "black leaf." Potassium deficiency can be common, especially on soils with high calcium or magnesium levels. There is high potassium uptake during bloom, and when this process is combined with abnormally dry conditions, insufficient potassium may be absorbed.

▲ ▲ Soluble fertilizers such as nitrogen can be easily injected into drip irrigation systems with very simple siphon injectors or more sophisticated equipment.

▶ Injecting fertilizers through the drip system can help provide vines with needed nutrients even in times with little or no rainfall.

However, deficiencies are seen only after véraison, when there is little a grower can do to compensate. Excessive potassium levels can lead to wine instability because of elevated juice pH and can also cause magnesium deficiency.

Potassium moves rather slowly in soils, and additions are typically made in a three- or four-foot band application underneath the trellis. Applications are most typically accomplished in the fall, so that seasonal winter and spring rains can move the potassium into the root zone, where it can be absorbed by the roots. Potassium fertilizers are typically applied in the form of potassium sulfate or, in some locations, potassium chloride. Banding (as opposed to broadcasting) is utilized in order to supersaturate the soil that has a heavy concentration of roots so that levels are high in relation to calcium and magnesium, thus allowing potassium to be more available in the soil solution.

There are numerous foliar fertilizer products on the market that contain potassium and whose manufacturers ensure optimal fruit maturity through readily available potassium supplementation. However, potassium is a macronutrient, and when actual levels of potassium contained in these products are compared with the amounts annually used by grapevines, the therapeutic value of these products becomes highly questionable.

Magnesium

Magnesium (Mg) is necessary in plants for the formation of chlorophyll, which helps produce energy in the vine. Magnesium also serves as a cofactor in enzymatic processes and is needed for nucleotide stability. Vines deficient in magnesium exhibit a characteristic interveinal yellowing of older, basal foliage, which typically appears sometime after fruit set. As deficiency worsens, basal foliage can abscise and the interveinal chlorosis can progress up the shoot. The damage to the vine from magnesium deficiency is due to reduced leaf area and a resulting decrease in photosynthesis.

Magnesium deficiency can be worse in growing seasons with a wet bloom to post-bloom period, which favors potassium uptake. In some grape varieties with pubescent rather than glossy foliage, magnesium deficiency can be corrected with sprays of magnesium sulfate (Epsom salts). The waxy cuticle of glabrous foliage inhibits uptake, but this obstacle may be overcome with additions of small amounts of nitrogen to the Epsom salt spray solution or through the use of specifically designed

Magnesium deficiency typically causes mild to severe interveinal chlorosis in basal foliage. In extreme situations, tissue can become necrotic and basal leaves may abscise.

magnesium-containing foliar nutrients. In low-pH soils, magnesium deficiency is commonly addressed through the application of dolomitic limestone. On higher pH soils, long-term treatment of magnesium deficiency is typically addressed with banded applications of potassium/magnesium sulfate in the fall.

Phosphorus

Phosphorus (P) is an element needed for energy transport within plants. It is an integral part of nucleic acids, and it is necessary for the conversion of starch to sugars and sugars to starch. Phosphorus deficiency is almost always confined to low-pH soils when phosphorus can combine with other elements, making it unavailable to vines. Phosphorus is highly mobile within vines, and deficiency symptoms appear as reddening between leaf veins on red-fruited varieties. Phosphorus uptake can vary by rootstock, and some scion varieties appear to exhibit deficiency symptoms at higher leaf concentrations than others.

Phosphorus moves very slowly in the soil, so if a pre-plant soil analysis reveals insufficient phosphorus, it should be applied and incorporated at a depth of four to six inches into the soil profile prior to planting. Phosphorus is usually added through the use of superphosphate fertilizers or in phosphoric compounds containing nitrogen. If a need for phosphorus is determined after a vineyard has been planted, applications should be banded at the edge of the weed-free zone under the trellis and lightly tilled to facilitate movement to grapevine feeder roots. When phosphorus is applied to high-pH soils, the application may exacerbate other existing micronutrient deficiencies, as of iron and zinc, for example.

Iron

Iron (Fe) is very immobile in plants and typically unavailable in the alkaline soils encountered in much of the Hill Country, West Texas, and the High Plains. There is plenty of iron in these soils, but it is chemically unavailable to many plants because of the high soil pH. Iron is important in plant respiration and photosynthesis, is an activator of enzymatic reactions, and is essential for chlorophyll synthesis. Because iron is immobile in plants, deficiency symptoms are expressed as interveinal chlorosis in new growth. In severe cases, chlorosis can progress to the point that foliage is absolutely devoid of chlorophyll (thus appearing white) and eventually necrotic, terminating the developing shoot. In many soils, the choice of a grape rootstock with *Vitis berlandieri* parentage, which is native to parts of Texas with high-pH soils, will be sufficient to overcome relative iron unavailability. In other situations, the injection of a mild acid into the irrigation water (to the point that water pH at the emitter is between 5.5 and 6.0) can improve iron availability for the roots in the immediate vicinity of the irrigation emitter.

Iron deficiency is exacerbated by excessively high or low water status. In either case, it is lack of root function that is responsible for the vine not being able to adequately forage for iron. Additions of phosphorous fertilizers can also worsen iron deficiencies. Because of iron unavailability in high-pH soils, the addition of elemental iron fertilizers does not correct iron deficiency. In these situations, the additional iron is immediately chemically tied up and is not available for uptake. While foliar fertilizers can provide very short-term correction of iron deficiency, these sprays only affect existing foliage, so new growth continues to express deficiency because of the immobility of iron in plants.

The single greatest tool in correcting iron deficiency is the application of iron chelate fertilizers. These chelated compounds resist being bound in the soil and thus remain available for root uptake. Such products are expensive but can provide season-long correction of iron deficiencies. As vines become mature, their root systems are more extensive and can

Iron is immobile in vines, so deficiency results in interveinal chlorosis of new growth rather than older tissue. In extreme cases, leaves can become completely devoid of chlorophyll.

thus better forage for iron, so iron deficiency symptoms are commonly less prevalent on older vines. There are numerous forms of iron chelate on the market, so be sure to discuss product types, rates, and timing of application with your local viticulture advisor should chelates need to be employed.

Zinc

As is the case with iron, zinc (Zn) deficiency is common on higher pH vineyard sites because the zinc ions are chemically bound to the soil. Zinc is an important component of plant enzymatic reactions and plant growth regulators, especially those governing apical dominance. Like iron, zinc is immobile in plants, and deficiencies are first seen as interveinal chlorosis in newer foliage. Other symptoms include zigzag shoot growth, a proliferation of lateral budbreak on shoots, and, in advanced cases, poor fruit set.

While zinc uptake can be enhanced with appropriate rootstocks, zinc deficiencies are normally corrected with zinc sprays applied very early in the season and the injection of zinc chelates into the drip system. If foliar sprays are to be employed, application must begin before newly emerging shoots reach one inch in length and must be continued biweekly through bloom. Soil applications are typically supplied through the drip system beginning in fall, with a second application immediately before budbreak.

Ample soil moisture during dormancy is important in zinc absorption, so additional water may need to be applied during dry winters.

Boron

Boron (B) deficiencies can occur across a wide variety of soil types and soil pH ranges. Boron deficiency is commonly one of hidden hunger, in which deficiencies are not suspected until cropping levels are adversely affected. Boron is important in the fertilization of flowers, and deficiency in grapevines can be seen in erratic fruit set, with some large fertilized berries and many much smaller, unfertilized berries. This condition is commonly referred to as "hens and chicks."

Zinc deficiency results in small, misshapen new leaves, shoots with a zigzag growth pattern, and a proliferation of lateral shoot growth.

While boron deficiencies are most common on soils with low boron levels, drought can greatly affect boron uptake. Under normal conditions, the flush of root growth that occurs during fall and early winter will facilitate boron uptake the following spring. Excessively dry autumn and winter conditions can reduce boron levels and impact yields in vineyards that otherwise perform normally. Boron levels are best monitored by late season petiole analysis so that action can be taken to prevent significant losses the following season. If boron deficiency is very mild, a single application of boron in the immediate pre-bloom spray should prevent poor set. If deficiency is moderate there should be two applications, one two weeks before anticipated bloom and one immediately before bloom. If boron deficiency is severe, fall soil-banded applications of boron should be made, as well as two spring foliar sprays. Exercise caution with boron additions because excessive applications can be phytotoxic.

Determining Nutrient Needs

It is important for new growers to thoroughly assess the potential nutrient profile of a new vineyard site and apply any needed amendments before vines are planted. In addition, timely monitoring of soils and plant tissue can help growers predict needed fertilizer inputs before deficiencies are found in vines. Sound sampling strategies and careful record-keeping can be crucial components of maintaining sound nutritional program in vineyards.

Taking a Soil Sample

Soil testing is a good first step in learning about the chemical nature of the media in which you will be attempting to grow grapevines. Although soil analysis has its limitations, it is an essential part of an optimal vineyard fertilization plan. Soil tests can advise a grower of "what is in the pot." In other words, it will indicate what base elements are present, what restrictive circumstances may result, and how those circumstances may affect the uptake of sufficient nutrients by a perennial crop.

The ability of a grapevine to extract sufficient nutrients and grow properly can be complicated by numerous factors, such as weed competition, overcropping, soil depth and drainage, improper irrigation, or inappropriate rootstock selection. Simply adding more nutrients under these situations may not remedy the problem. Grapevines also respond differently on the same site in different years due to differences in temperature and rainfall and available sunlight. Consequently, for perennial crops, soil sampling alone will not allow a grower to fine-tune a vineyard fertilization program; tissue testing must accompany it. Taking soil samples prior to planting and every five to six years in the life of a vineyard can provide additional insight into grapevine responses to fertilizers over time.

To collect a soil sample, choose several random locations in the general area of the vineyard. Ideally, both a topsoil and a subsoil sample will be collected, at least with the initial pre-plant soil testing. With a soil borer or narrow blade shovel, collect slices from the soil surface to eight inches deep at each location. Together, these will represent the topsoil sample. Continue digging in each location and collect a second set of samples from eight to at least fifteen and preferably twenty inches deep. Keep these soil slices separate from the shallower set of slices, as the deeper ones represent the subsoil sample. Submit these samples to a reputable soil testing lab for analysis. Remember, many of the recommendations that are autogenerated by the soil testing lab are generally geared to annual field crops. These computer models may not take many subtle variables into account. Review the results with your local viticulture advisor or professional consultant before making nutrient applications.

Petiole Testing

For grapevines, petiole (leaf stem) tissue is the part of the vine that is most reflective of soil/plant/seasonal interaction. Since there is variable mobility of nutrients within the vine during the season, two distinct sampling periods have been identified that will accurately reflect nutritional status, thus allowing growers to accurately supply fertilizers to meet vine needs. Sampling periods are either at 50 percent bloom or at least seventy days post-bloom (thus, between véraison and harvest). Sampling at other times may help identify deficiencies, but those results may not be an accurate or quantitative predictor of grapevine needs because nutrients are in a rapid state of flux or mobility.

Routine petiole sampling is best done annually but should be done at least every other year in order to track the performance of vines and the response to fertilizer inputs. Many growers choose to sample the same "sentinel" vines in a given block every year as an indicator of whole block response. Likewise, some growers sample half of their vineyard blocks one year and the other half the following year, which gives them insight as to seasonal effects on the vineyard as a whole. Sampling should be done on blocks of the same variety and age and on the same rootstock. If blocks are planted on more than one distinct soil type, those blocks should be sampled separately.

Petiole sampling is the best way to determine the nutritional needs of grapevines.

If a problem arises in the vineyard and nutritional problems are suspected, sampling can be done at any time, on both affected and nonaffected tissue. Although the numbers provided in the results are empirically meaningless, differing nutrient levels between samples may shed light on the problem.

Sampling at 50 percent bloom. In western US growing regions sampling is commonly done at the point of 50 percent bloom, which is thought to be the single best time to analyze nitrogen levels in vines. This point in the season is also considered to be optimal for the analysis of zinc, although the application of zinc after bloom has little impact on vine performance that season. To sample during this time, select two primary bearing shoots, well exposed to sunlight, on each of twenty vines in a given block. Sample the leaf that is subtending (opposite from) the basal fruit cluster on each shoot. Remove and discard the leaf blades and retain the petioles. These forty petioles will serve as a representative sample of the block.

Sampling at seventy days post-bloom. Sampling at seventy (or more) days post-bloom is common in many eastern US growing regions and is considered the best period in which to determine the magnesium and potassium needs of grapevines for the following season. Likewise, boron levels may indicate the need for fall soil applications or pre-bloom fo-liar applications the following season. Samples can be collected as early as seventy days after full bloom (véraison), but many growers choose to sample right before harvest. Nelson Shaulis maintained that post-harvest sampling can also be employed later in the fall as long as the canopy has not begun to senesce. To sample during this period, once again collect from two primary bearing shoots, well exposed to sunlight, on each of thirty vines within a block. Rather than choosing leaves subtending fruit, move farther down the shoot to the most recently matured leaf that has reached full size. Again, remove the leaf and discard the leaf blade, retain-ing sixty petioles for the sample.

Sample preparation. Petioles should be gently washed to remove ad-sorbed (surface retained) nutrients as opposed to absorbed (internally ac-tive) nutrients. Dust and chemical residue can result in spurious analysis results if the petioles are not properly washed and dried. Prepare a warm water bath with a small amount of detergent that does not contain phos-phorus. Dreft® brand diaper detergent or similar products are commonly used by growers, but many analytical labs will use 1 percent hydrochloric acid if they are preparing a petiole bath. Dunk each forty-petiole sample in the bath and rub gently for about fifteen seconds. Remove and place the sample into a separate bath of distilled water and rinse gently. Re-move and transfer to a second, then a third bath of distilled water and dry the sample on a clean paper towel. Replace washing and rinsing baths after three or four samples to keep cross-contamination to a minimum.

After petioles have been washed and dried, transfer them to an open small paper bag and leave them in a warm to mildly hot environment. Laboratories use warm forced-air dryers, but a building attic works just fine. After petioles are completely dry, they are ready for submission to a lab for analysis. Label each sample bag with at least your name and block identification number and include a sheet listing the identity of all of your samples.

Interpretation of test results. The specific characteristics of a growing season can have a tremendous impact on the results of petiole analysis, making interpretation of lab findings as much of an art as a science. For example, extended drought may greatly reduce root growth and nutrient uptake, resulting in low boron levels in the fall, but in such a case the real need is for rainfall, not more boron in the soil. In addition, while many labs are quite capable of conducting sound analysis, growers should be

Table 16.1. Target petiole nutrient values

Nutrient	Chemical symbol	Target petiole values at bloom	Target petiole values at véraison/harvest
Nitrogen	N	1.2–2.5%	0.8–1.4%
Phosphorus	P	0.15–0.4%	0.1–0.3%
Potassium	K	1.5–3.0%	1.5–3.0%
Calcium	Ca	1.2–3.0%	1.0–3.0%
Magnesium	Mg	0.5–0.75%	0.5–1.0%
Iron	Fe	30–100 ppm	30–100 ppm
Zinc	Zn	30–100 ppm	30–100 ppm
Manganese	Mn	25–1,000 ppm	100–1,000 ppm
Boron	B	25–100 ppm	30–100 ppm
Copper	Cu	6–25 ppm	6–25 ppm
Molybdenum	Mo	~0.5 ppm	~0.5 ppm
Sodium	Na	< 1,000 ppm	< 1,000 ppm

cautious of who interprets the results and recommends fertilization practices based on those interpretations. Most nutrient analysis labs commonly make recommendations for field crops, but their understanding of perennial crops may be limited. Be sure to discuss the petiole analysis results with a trusted viticulture advisor or consultant before acting on the recommendations of a lab.

Nutrient Availability and Competition

The pH level of a soil can affect nutrient availability and the ability of a vine to grow. Soil pH is a logarithmic description of the relative acidity or alkalinity of a soil. A pH of 7 indicates that there are equal amounts of hydrogen (acid) or hydroxyl (alkaline) ions in the soil solution, and the rise or fall of one unit indicates a tenfold increase in one of these ions.

Higher pH soils are more alkaline, and lower pH soils are more acidic. In acidic soils, aluminum is available to grapevines to the point of aluminum toxicity for *vinifera* and some hybrid varieties. American varieties such as 'Concord' evolved in low-pH soils and do not show a sensitivity to increased aluminum availability. The *vinifera* varieties are more adapted to neutral or alkaline soils, so the soil pH should be raised to at least 6.0 to avoid problems. Applying agricultural lime to a vineyard site can remedy a soil that is too acid for a particular variety, and the type of lime (calcitic or dolomitic) and amount will be dictated by relative nutrient levels revealed by the soil analysis.

Some elements compete with each other for uptake by grapevines. The most critical balance in a vineyard is generally the competition between calcium, magnesium, and potassium. An overabundance of any of these elements can inhibit the uptake of the others. For much of the western part of Texas, the high calcium bicarbonate content of the predominantly alkaline soils restricts the availability of magnesium and potassium. This restriction may result in magnesium deficiency symptoms in vines of all ages, and as a vineyard comes into production, potassium availability and uptake may also be affected. Occasionally, one will see agricultural analytical labs prescribing specific balances of nutrient levels based on percentage base saturation between these elements. While this approach may be sound in principle, there are a number of factors, such as soil depth, texture, and cation exchange capacity, that complicate this issue. Petiole testing is the single best way to determine these balances and to predict the potassium and magnesium needs of a vineyard.

Another competitive relationship that commonly affects vineyards is that between phosphorus and both iron and zinc. An overabundance of phosphorus, and more specifically, the addition of fertilizers containing phosphorus, can cause or exacerbate deficiencies of iron and zinc. In alkaline soils, phosphorus deficiencies are rare, and caution should be exercised when contemplating the addition of this element.

Diseases Affecting Foliage and Fruit

O NE OF THE MOST daunting aspects of growing high-quality wine grapes is the management of the myriad fungal pathogens that can affect both foliage and fruit. In many parts of Texas, growers can produce clean fruit with little or no insecticide inputs, but fungal pathogens are active in all areas of the state and require a well thought out management plan that includes conventional fungicides. Many have tried organic approaches to managing fungal pathogens, but organic options provide marginal control at best and some "organic" materials in fact have a far more negative impact on the environment than synthetic fungicides.

There is no recipe for "clean" fruit; nobody can write out a single spray schedule that will cover all varieties in all areas under any set of weather conditions. It is incumbent upon all growers to understand the susceptibility of the varieties they grow and to fully comprehend the disease cycle of all pertinent fungal pathogens in order to successfully manage this complex threat to producing high-quality fruit. This brief overview is intended to be a starting point for understanding the wide range of disease threats and is by no means intended to replace the in-depth understanding of disease epidemiology required to make informed management decisions. In addition, there is tremendous variability in disease susceptibility among varieties, and there are strong regional and seasonal conditions that influence disease pressure. Fungal management guides prepared by Texas A&M AgriLife Extension and other state land grant institutions provide information on fungicide selection and timing. The American Phytopathological Society produces an array of disease compendia, and *The Compendium of Grapevine Diseases* is an excellent guide to the identification and understanding of grapevine diseases that occur worldwide.

Anthracnose

Caused by the fungal pathogen *Elsinoë ampelina,* anthracnose is also known as bird's eye rot because the fungal lesions on fruit resemble a bird's eye. Before downy mildew and powdery mildew were introduced into Europe, anthracnose was the most damaging disease of grapes and was only partially controlled after the introduction of Bordeaux mixture (a fungicide composed of copper sulfate, hydrated lime, and water) in the late nineteenth century.

The disease overwinters as elliptical lesions on canes, and these lesions give rise to infections on all green tissue—shoots, leaves, and young fruit. On basal foliage, anthracnose produces small round to angular lesions with brown to black margins. Infections on tender shoots appear as small elliptical lesions with purplish-black margins and may cause necrosis deep into the shoot tissue, rendering the shoots brittle and subject to breakage. Fruit infections first appear as purplish lesions that develop a black ring surrounding the infection and progress to a grayish-white color. Fruit is subject to infection from bloom up to véraison.

While most varieties are relatively susceptible to anthracnose, the disease is normally controlled by fungicide spray programs designed to manage other diseases. 'Blanc DuBois,' however, is extremely susceptible to anthracnose and typically requires additional measures to successfully prevent significant damage. Sprays of limed sulfur (calcium polysulfide) are routinely applied to 'Blanc DuBois' vineyards after pruning toward the end of dormancy to neutralize cane lesions and reduce inoculum for subsequent infection. Conventional fungicides are sometimes needed during the growing season to fully control this disease.

Black Rot

Black rot (*Guignardia bidwellii*) is potentially the most economically damaging disease to commercial grape production across North, Central, and East Texas. While black rot has been found on the High Plains and in West Texas, arid conditions limit potential infection in all but abnormally wet growing seasons.

Black rot overwinters primarily as mummified infected fruit, either on the ground or retained in the canopy. A single mummified berry can give rise to thousands of airborne spores that are capable of infecting a wide area within a vineyard. Primary infection takes place starting at about ten-inch shoot growth, but discharge of ascospores from mummies peaks over a month-long period, with bloom and fruit set

Cane lesions of anthracnose. 'Blanc DuBois' is especially susceptible to this disease, and control measures usually begin with the application of limed sulfur (calcium polysulfide) during dormancy.

Fruit infections from anthracnose, commonly referred to as bird's eye rot.

occurring at the center of this disease pressure period. Primary infections on leaves and shoots subsequently threaten fruit for the remainder of the season. Although berries themselves become resistant to infection once they reach approximately five degrees Brix (5 percent soluble solids), the green rachis remains a susceptible target up to the point of harvest. Rachis infections cause a collapse of vascular tissue within the rachis, which results in a secession of sugar transport into fruit and fruit loss from breakage of the brittle rachis.

Primary infection typically appears as brownish angular lesions on basal foliage seven to ten days after infection. These lesions over time develop a black margin around the edge. A few days after the initial lesion appearance, black pycnidia form within the lesion and serve as an inoculum source for secondary infection. Cane, petiole, and rachis infections

Black rot infection of leaves and fruit. This disease is the Achilles' heel of organic grape production.

appear as black lesions that become elliptical in shape, with pycnidia also forming within these infection sites. Fruit typically begins showing symptoms of infection with the formation of a reddish-brown ring that quickly engulfs the entire berry. Within a few days, the berry begins to wither and dehydrate into a mummified berry, which serves as a potent source of further disease development. Once a black rot fruit infection has been established, it can be virtually impossible to prevent significant crop loss.

As with most fungal pathogens, black rot infection is driven by temperature and wet conditions. Infection occurs not as a consequence of how much rain falls but rather how long the canopy remains wet. Canopy management can play a large role in inhibiting or expediting canopy drying after a rainfall event. Bob Spotts from Ohio State University developed a chart to model the temperature and duration of leaf wetness necessary for black rot infection to develop.

This model can be extremely useful to growers because knowing when conditions are optimum for infection can provide guidance as to when to apply fungicides that have post-infection activity. Post-infection does not mean post-symptom activity. It normally takes seven to ten days from the beginning of an infection period until the time symptoms appear on

Table 17.1.
Environmental conditions necessary for primary infection of black rot

Average temperature	(°F)Hours of leaf wetness
< 50	No infection
50	24
55	12
60	9
65	8
70	7
75	7
80	6
85	9
90	12
> 90	No infection

leaves, shoots, or fruit. Applying fungicides after symptom development will not affect the damage that has already been done.

Black rot is the Achilles' heel of organic grape production because neither sulfur nor copper has any efficacy in controlling this disease. While spray programs that do not cover black rot may be successful in the arid portions of the state, successfully managing this disease will be essential for producing high-quality grapes in most regions.

Downy Mildew

Plasmopara viticola, the fungal pathogen that causes downy mildew, can be every bit as devastating to a grape crop as black rot. Downy mildew leaf infections can result in vine defoliation during the middle of the growing season, preventing the crop from properly maturing and having a negative impact on the health of grapevines. This defoliation is problematic because the loss of photosynthetic leaf area prevents the vines from accumulating the reserves necessary to initiate growth in the next season.

Grapevines suffering from defoliation due to downy mildew infections commonly have buds with reduced winter hardiness. Although downy mildew is considered to be a disease of grapevines in warm, wet climates, it is problematic in every major grape-growing region of the world. California is generally thought to not have downy mildew, but evidence suggests that it survives at a very low, insignificant level and becomes problematic only in years with unseasonable rainfall.

▲ Downy mildew leaf lesions. Yellow discoloration can be seen on the upper surface of leaves, but fungal sporulation occurs on the lower surface.

▶ Active downy mildew infection and sporulation on fruit.

The disease overwinters in fallen leaf litter and in the soil, and primary infection occurs in the spring or summer with splashing rainfall and wet, dark conditions, such as those provided by late afternoon or evening showers. Newly infected leaves begin to exhibit yellowish lesions with an oily, sunken appearance, and in a few days sporulation begins on the lower side of the leaf surface, providing sufficient inoculum for further disease development. Infected shoot tips develop a tell tale shepherd's crook

Downy mildew infection of the rachis becomes systemic and affects the ability of fruit to ripen. Once systemic, these infections cannot be controlled.

appearance with a proliferation of white spores. Once leaves become infected, secondary infection may spread to the fruit very quickly, leading to significant economic loss. Secondary infection is not dependent on dark conditions. Infections in fruit or rachis tissue become systemic and are no longer controllable. Infected fruit fail to achieve acceptable maturity and, if harvested, can cause detectable defects in wine quality.

While this disease was traditionally controlled with Bordeaux mixture in Europe during the nineteenth century, there are now numerous modern fungicides that are specifically labeled as effective in preventing or curing downy mildew infections. In some seasons when infection occurs during intervals between sprays, the heat and intense sunlight of summer can come to the rescue and burn out existing leaf lesions. Growers should not become complacent and dependent on this phenomenon, however, because a single summer thundershower can cause new rounds of infections and tropical moisture can become a catastrophic event. Should tropical weather patterns approach, the best line of defense is to apply a full rate of a rainfast fungicide and be prepared to follow up to treat post-infection activity.

Phomopsis Cane and Leaf Spot

Although it had probably been a problem for quite some time, the presence of phomopsis cane and leaf spot (*Phomopsis viticola*) was first confirmed in Texas in 2005. A cool, wet season disease, phomopsis in Texas vineyards causes symptoms that are much more subtle than those that appear in the northeastern United States.

Phomopsis overwinters as lesions on one-year canes and older permanent wood structures of the vines, such as spurs and cordons. The disease cycle begins as early as one-inch shoot growth, when leaves and shoots are subject to infection. Economic damage is most likely to occur at the three- to five-inch shoot growth stage when emerging clusters first become visible. Infection at this stage can lead to the loss of cluster wings or entire clusters. Infection first appears as blackish elliptical lesions on shoots and rachises and as subtle spots on leaves. These lesions can coalesce and become necrotic, typically on the first few basal leaves of a shoot.

Economic damage occurs with infection of the cluster rachis or through direct berry infection. Early-season rachis infections appear similar to those on shoots, but as fruit maturity approaches these infection sites become necrotic, impeding translocation of photosynthates to the fruit and causing the cluster itself to become brittle. Because of the brittle nature of infected rachis tissue, this disease becomes more problematic when growers attempt to prolong fruit hang time to maximize the soluble solids content of the fruit.

Effective disease control begins very early in the season, with fungicide spray applications starting at three-inch shoot growth or earlier in the presence of high disease pressure. Because grapevines grow very rapidly early in the season, it is challenging for growers to keep fungicide coverage on expanding and newly developing tissue. Typically, the frequency of fungicide application is driven by previous disease pressure and the amount of rainfall in a given season. With spring shoot expansion, fungicide applications become more important once rachis and flower cluster tissues become visible. The period from immediate pre-bloom through fruit set is when rachis and berry tissues are most susceptible to infection. Early-season infections are the most critical for the rachis and can result in the loss of entire clusters.

▲ Lesions of phomopsis cane and leaf spot on green shoots.

◀ Lesions of phomopsis cane and leaf spot on overwintering dormant canes.

Phomopsis rachis infection occurring near bloom becomes active near fruit maturity when rachis tissue collapses, inhibiting sugar accumulation and causing fruit to drop prematurely.

Powdery Mildew

Perhaps the most prevalent fungal infection of grapes around the world, powdery mildew is a disease that is seen every year on susceptible varieties, but prophylactic sprays keep the disease at bay, thus mitigating negative impacts on yield, wine quality, and vine health. Caused by the pathogen *Erysiphe necator*, powdery mildew overwinters as spores on buds but more importantly as cleistothecia that mature in late summer and fall and appear as red blotchy spots on mature year-old canes. In the spring, ascospores are released from these cleistothecia, with temperatures as low as 50 degrees Fahrenheit and thirteen hours of canopy wetness giving rise to primary infection. With moderate temperatures, no further rainfall is needed for secondary spread, and if left unchecked, powdery mildew can quickly cause widespread damage to fruit and foliage.

Management of powdery mildew usually begins at twelve-inch shoot growth, with the inclusion of contact fungicides such as sulfur or oil in combination with conventional fungicides used for the management of other fungal diseases. On the West Coast, powdery mildew spray application begins at budbreak, and in the High Plains and West Texas early applications may play an important role in disease management. Although the majority of the powdery mildew fungus appears on the surface of leaves and fruit and can be readily killed by contact fungicides, it is unwise to limit control efforts to eradicating existing infections because significant leaf and fruit injury can occur. More potent fungicides with locally systemic attributes for either preventive or post-infection activity are utilized from the pre-bloom stage until about four to five weeks after bloom. At that point, berries become resistant to infection, but the disease can continue to be problematic on the leaves, shoots, and, more importantly, on the rachis, where it can lead to cluster collapse. Severe powdery mildew infections can cause direct loss of fruit from cracking, and as little as 2 percent berry infection can be detected as a wine defect by tasting panels.

Although there are currently a number of very effective fungicides commonly used to control powdery mildew, all of these synthetic materials may promote the development of resistant races of powdery mildew fungi. Compounds such as sulfur, copper, and oils are not thought to promote resistance. Chemical selection, timing, and rotation are all important components of a resistance management strategy that will allow these fungicides to remain effective into the future.

The biggest enemy of powdery mildew is sunlight, and infection is much more prevalent in shaded areas of the canopy. Consequently, shoot positioning and canopy management techniques that expose fruit and basal leaves to sunlight can play a significant role in reducing disease pressure on fruit and foliage.

Powdery mildew injures foliage by reducing photosynthesis, and the amount of damage is proportional to the percentage of foliage infected. Unchecked infections can cause premature vine defoliation leading to reduced cane maturity and a loss of vine winter hardiness. Growers are challenged in many parts of the state in managing powdery mildew several months after harvest. Once fruit is no longer present, the materials of choice for powdery mildew control return to sulfur, oils, and other contact-type fungicides that are not prone to resistance.

Botrytis and Summer Fruit Rot Complex

There are a number of pathogens that can cause widespread fruit loss just as fruit begins to ripen. It is important to distinguish between these organisms because fungicides that may control one pathogen can be entirely ineffective against many others. While present worldwide,

Powdery mildew infection on foliage and fruit. Severe infection can cause fruit cracking and vine defoliation. Even mild infections negatively affect fruit quality and the photosynthetic efficiency of foliage.

Botrytis cinerea is the most problematic rot organism at or just before harvest in the hot, humid portion of the southeastern United States, but other rot organisms, commonly called the summer rot complex, are far more common. Grape varieties that grow in tight clusters or environmental conditions that cause tight clusters can drastically increase the incidence of the disease and severity of bunch rot problems. Leaf removal can help remove abscised calyptra and floral debris that can be a contributing factor to subsequent berry infection.

Botrytis infection can occur at bloom, though lesions remain latent until fruit begins to ripen. In very wet years, *Botrytis* can infect leaves and shoot tips, but sprays are generally directed at controlling fruit infections. On susceptible varieties, specific fungicide sprays are also timed right before bunch closure (three to four weeks after set, as the berries swell and begin to touch), when spray materials can still contact tender tissue. On some varieties, such as 'Pinot Gris,' bunches are so tight that the final expansion of berries results in some berries being separated from the fruit stem or pedicel, resulting in a release of juice inside the cluster. Because soluble solid concentration of the fruit is relatively high at this point, the juice makes an ideal substrate for fungal growth. Under frequent or heavy rainfall at this time, further fungal development and fruit loss may be inevitable. Obviously, selecting appropriate varieties for a specific climate are important factors in preventing these losses. There are many *Botrytis*-specific fungicides that can be applied from véraison to maturity, but most of these are active only against *Botrytis* and not other rot pathogens.

Besides *Botrytis* infections, there are a wide variety of other fungal and bacterial rot organisms that can cause losses of yields and fruit quality. The summer rot complex most commonly comprises ripe rot (*Colletotrichum* spp.), bitter rot (*Greeneria uvicola*), Macrophoma rot (*Botryosphaeria dothidea*), and the sour rot organisms (*Alternaria, Aspergillis, Cladosporium, Rhizopus, Penicillium,* and others).

Heavy rainfall at or near harvest may make the presence of one or more of these organisms a foregone conclusion, but growers can take

Fruit collapsing from any one of a number of summer rot fungal pathogens. Black sporulation in fruit suggests that *Aspergillis* is at least one of the culprits.

precautions that may limit losses. While some consultants favor withholding water from grapevines for the first month after bloom to reduce berry size and maximize wine flavor potential, tropical rains that come later can cause these small berries to split, providing a prime entry site for opportunistic decay organisms. With this situation in mind, growers can also help reduce summer rot problems by altering irrigation strategies.

Other opportunities for rots to gain a foothold in clusters include hail injury, insect feeding, and bird damage. Sanitation procedures such as removing old rachis and cluster material from the canopy during winter pruning can reduce both disease inoculum in the fruiting zone and, subsequently, future disease incidence. While there are a few fungicides that specifically target this group of pathogens, most growers believe that applications of broad-spectrum fungicides such as Captan for downy mildew control can also aid in summer rot control when harvest seasons become wet.

Leaf Blight

Also known as Isariopsis leaf spot, leaf blight has been reported as occurring primarily on wild grape species. Although reports of this disease causing economic losses in vineyards primarily cite tropical and subtropical locations, this fungal pathogen has caused crop failure in a few Texas vineyards. Lesions appear on leaf surfaces as angular brown infections between two and twenty millimeters in diameter and may coalesce under severe infection. Fruit infection appears very much like black rot, but whole clusters generally collapse.

Because this disease is so uncommon, fungicide labels rarely list activity against this pathogen, although it appears that both strobilurin and sterol-inhibiting fungicides have some activity. At this writing, leaf blight has been confirmed in Texas only in Hill Country vineyards, but it probably exists in other growing areas of the state as well. 'Cabernet Sauvignon' appears to be highly sensitive to this disease.

Fungal lesion of leaf blight on 'Pinot Blanc.' While not widespread at this time, this fungal pathogen can cause severe crop loss under wet summer conditions.

Other Fungal Pathogens

There are numerous other opportunistic fungal pathogens that can appear on grape foliage, but in many cases such pathogens do not warrant fungicide treatment. An example of one of these organisms is *Septoria ampelina,* causal organism of Septoria leaf spot. This mild foliar infection is common late in the season, especially in years with late-summer and autumn precipitation. In most cases this disease is of little consequence and is rarely if ever treated.

Managing Fungal Diseases

For growers wanting to establish a sustainable *commercial* vineyard in Texas, the use of conventional fungicides will have to be a concept they are comfortable embracing. No amount of personal will can prevent these pathogens from finding your vines and freely infecting them when the weather provides opportunity. For backyard production in which a crop loss will not lead to economic hardship, growers are free to explore and employ organic materials and cultural practices that will reduce infection potential.

In addition to copper and sulfur, which are considered organic materials, there are numerous other products on the market that aim to trigger an immune response in vines and thus prevent infection. This process

Septoria leaf spot, one of several chronic fungal pathogens that seldom require treatment.

is known as systemically acquired resistance (SAR) but should be viewed only as a modestly effective tool in combating these pathogens.

Commercial growers need to become familiar with the vast array of labeled fungicide materials. When choosing fungicides, look carefully at their efficacy against specific diseases, how they work , the federally mandated intervals from the time of application until workers can reenter the vineyard or the fruit can be harvested, and techniques needed to manage resistance of pathogens to these important tools. Because fungicide efficacy, labeling, and availability can vary considerably over time, discussion of specific fungicide products or chemistries has been avoided in this text. Refer to the latest disease control guide offered by your extension specialists or to guides from a similar production region.

For commercial growers, an airblast sprayer will be a necessity. Hydraulic sprayers may be able to deliver material to the outer surface of a vineyard canopy, but airblast sprayers utilize a strong blast of air to move leaves and help deposit material far into the interior of a canopy.

One of the most important components of effectual disease management is thorough coverage of the canopy. While some growers achieve adequate disease control using low volumes of water per acre in their spray program, this technique is best reserved for early in the season, when there is less leaf tissue to cover. Once a full canopy has developed, low water volume applications may not adequately cover foliage and fruit and may fail to control disease. When spending money on fungicides, it only makes sense to take the time to apply them with sufficient water volume to optimize their effectiveness. An excellent reference book, *Effective Vineyard Spraying: A Practical Guide for Growers,* by Andrew Landers of Cornell University, is highly recommended for new and seasoned growers alike.

Managing fungal pathogens involves timely spraying aimed at disease prevention rather than trying to stop a roaring epidemic after it gets established. Many of these diseases are most active from two weeks before bloom to approximately a month after bloom, and the most effective fungicides are commonly applied at these times. Monitoring weather conditions and predicting fungal infection periods can be a tremendous means of improving disease control and reducing fungicide costs. Weather stations that monitor temperature, rainfall, and leaf wetness can also be a wise investment,

Other diseases, such as phomopsis, are most effectively managed by early-season spraying, with additional applications at or after bloom. It is not uncommon for a grower to effectively prevent fungal diseases early in the season, go into midsummer with a squeaky clean canopy and fruit load, then not have to worry about further disease infection while hot, dry, summer conditions prevail. In these cases, further spraying may not be needed until after harvest, when moderating temperatures and autumn precipitation may promote powdery and downy mildew pressure. For post-harvest sprays, less expensive contact fungicides are commonly employed, but it is extremely important to maintain a healthy canopy as long as possible through the fall to maximize post-harvest photosynthesis. Carbohydrates accumulated during the fall are important in optimizing grapevine cold hardiness and as a source of energy when the next growing season begins.

Weather stations can help growers make decisions on the management of fungal pathogens.

Other Diseases Affecting Grapevines

Crown Gall

A FTER PIERCE'S DISEASE, the second most important bacterial disease of grapevines is crown gall, caused by the pathogen *Agrobacterium vitis.* Similar galling diseases of other woody plants exist in nature, but the species that causes disease in grapevines is unique to the genus *Vitis.* It was previously believed that when grapevines suffered freeze injury, the pathogen entered those wounds from the soil and caused the

Typical gall at the graft union. Most vines that show crown gall symptoms came from the nursery already infected.

formation of tumorlike galls that interrupted the flow of water and nutrients through the vascular system of a vine. Research has now shown that most vines that exhibit crown gall symptoms were probably already infected and typically came that way from nurseries, but it is a cold weather event or other mechanical injury that triggers galling within the vine.

Once infected vines are planted in a vineyard, the site is contaminated and will re-infect healthy transplants. Galls are most commonly seen at the graft union or near the ground but can extend to more distal parts of the vine. Galls may be inconsequential or may eventually kill the entire vine. Crown gall lowers vigor and productivity of grapevines to varying degrees, but at some point infected vines are no longer profitable and should be removed.

To date, there is no chemical that is effective in controlling crown gall. Avoiding sites prone to hard freezes is the best way to limit crown gall problems, but in the Texas climate, there is no site that is not at risk from an injurious freeze event over time. Crown gall can be spread in the vineyard from vine to vine, so sanitation and removal of severely in-

fected plant tissue can help control the spread of this disease. The bacterium can reside in the soil, and, consequently, replanting in a space previously occupied by a vine infected with crown gall increases the probability that subsequent vine infection will take place.

The best way to avoid crown gall problems is to not replant on ground contaminated with *Agrobacterium* and to purchase plant material free from the bacterial pathogen, but doing so is difficult. At this point, even plant material that is "certified disease free" from the nursery may well be infected with crown gall. Thankfully,

Galling can also take place in more distal parts of the vine when tissue suffers freeze injury.

this situation is being addressed by the National Clean Plant Network, and sensitive new screening procedures for crown gall contamination have been developed. From these efforts, a consortia of nurseries is developing new mother blocks of plant material free from virus and crown gall, and these "clean" plants will be used for subsequent grapevine propagation.

Fungal Wood Pathogens

There are a number of fungal pathogens, some only recently discovered, that invade and kill or devitalize grapevines. The disease most commonly associated with this group is eutypa. While the *Eutypa* pathogen has been found in Texas, it is by far the least common among those that have caused fungal galling and vine decline. Many growers have viewed eutypa disease fact sheets stating that a wedge-shape dead area in the cross-section of a trunk is a definitive symptom of this disease. This statement is not entirely true, because that zonal death pattern is simply how grapevines die. Whether from freeze injury or pathogen invasion, wedge-shaped necrotic areas are common in trunks, especially in more aged vineyard blocks. Many of the fungal pathogens of grapevines can cause these same wedge-shaped dead zones, but they all form some type of initially subtle canker on cordons or trunks. After infection, vines begin a steady decline in growth and productivity, and some of these pathogens can kill grapevines outright. Other pathogens can simply render vines unproductive, forcing the removal of the vines. Points of infection can be identified because new green growth at and past the point of infection appears stunted, with variable distortion of shoots and foliage.

One of the oldest known diseases of grapevines is esca, or black measles. Since Roman times, it has been described as a disease that causes distinct foliar symptoms of chlorotic to necrotic interveinal scorching, which can cause premature leaf fall. Symptoms vary by variety and can be distinct one year and absent the next in the same vine. Affected fruit have purplish-black lesions, which can lead to fruit cracking before maturity. Several fungal organisms have been associated with this disease, but no single pathogen has yet been identified as being responsible for this disease. Esca can cause a decline of grapevines, leading to the death of cordons or whole vines.

From 2007 to 2009, a survey of forty-five Texas vineyards revealed eleven species of fungal canker organisms capable of infecting and causing the decline of grapevines. Among the most virulent of these

Typical foliar symptoms of esca. Sometimes confused with Pierce's disease leaf scorching, esca symptoms are far more regular within leaf tissue.

organisms were *Neofusicoccum parvum, Lasiodiplodia theobromae, Botryosphaeria dothidea, Phomopsis viticola* (the causal organism of phomopsis cane and leaf spot), and *Diplodia corticola.* The six other fungal species identified showed lesser ability to aggressively colonize grape tissue but can still be problematic in some situations. These fungal pathogens are collectively referred to as "bot canker." Prevention of and treatment for all of these infections is the same. These diseases do not move rapidly in grapevine tissue, so identification and removal of infected vines or cordons is critical to management of this disease complex. Pruning wounds, especially those greater than one-half inch in diameter, present the most problematic sites for new infections.

The practice of double-trunking vines is a good way to reduce potential losses from wood-rotting agents. If infection is seen in one trunk, it can be removed and cropping can be compensated by retaining a cane from the remaining trunk until a new second trunk can be developed. Delaying pruning until late winter can also aid in reducing infection from fungal wood-rotting pathogens. A pruning cut made in late fall may remain a viable site for infection for as long as five weeks. The same cut

Dead cordons and stunted, irregular shoot growth early in the season are symptomatic of infection from any one of a number of fungal wood canker diseases.

made at budbreak is susceptible to infection for only about five days because, once vines begin to grow, wound healing is greatly enhanced. While it is physically impossible to do all pruning at budbreak, it does make the case to not prune vines in late fall or early winter to get the task behind you.

When large pruning cuts are made on vines, it is wise to seal them with an asphalt-based pruning paint. Diluted water-based latex house paint can also provide a physical barrier to infection but may wash off with a rain shortly after application. *All wood suspected of being infected with these rot organisms should be removed from the vineyard and buried or burned.* Some fungicides have also received supplemental labeling for the control of fungal wood canker pathogens, and applications are recommended after pruning each day. Consult your local viticultural advisor for specific material and application rate selection for your area.

Leaf Roll Virus

The closely related virus pathogens collectively named leaf roll (or red leaf) are the most widely distributed virus pathogens of grapevines in Texas and have resulted in widespread loss of vine vigor and productivity. Leaf roll is most easily identified on red-fruited varieties, and symptoms begin with the red coloration of basal foliage in early summer. On white-fruited varieties, foliage becomes chlorotic while the leaf veins remain green.

Symptoms of leaf roll commonly begin appearing in the second or third year after vineyard planting, but first-leaf vines may show some symptoms as well. Various leaf roll pathovars cause slightly differing symptoms, but downward cupping of foliage is a common symptom among various strains. Leaf roll infection does not kill vines outright but is more of a chronic problem that will slow vegetative vigor and delay and reduce sugar accumulation in fruit. Leaf roll is also thought to reduce fruit color in red-fruited varieties and overall vine cold hardiness in all infected varieties.

Introduction of leaf roll virus to a new vineyard site is almost always the result of contaminated nursery stock. Many rootstock varieties do not show symptoms when ungrafted, but when they are grafted with a susceptible scion the disease may be expressed at some point in the life of the vine. Unlike crown gall, "certified disease free" vines from the nursery are theoretically free from the pathogen. Because leaf roll is so widely problematic, spending the extra money for the purchase of disease free nursery stock is a wise investment. Once introduced into a vineyard, the disease can be spread by a number of mealy bug species. An infestation is typically not noticed in a vineyard until mealy bug numbers climb and leaf roll symptoms begin to spread. Systemic neonicotenoid insecticides active against PD vectors will also control mealy bugs. As with most chronic diseases, at some point a grower will probably make the decision to replace infected vines.

Other Grapevine Diseases

There are numerous other viral diseases of grapevines, some of which are problematic and others that are inconsequential to *V. vinifera* varieties. Tomato ringspot virus and tobacco ringspot virus can cause vine decline in American and some French-American hybrid grape varieties, but these pathogens rarely produce symptoms or cause vine decline in most *vinifera* varieties even after infection takes place.

At this writing, a new geminivirus disease named red blotch has been identified in California, New York, Virginia, and Texas. While apparently this disease may not produce symptoms on varieties such as 'Black Spanish' or 'Blanc DuBois', it can have serious consequences for most *vinifera* varieties. Red blotch causes the formation of irregular red or pink veins and leaf blotches within green leaves during late summer, and it greatly slows and reduces sugar accumulation in the fruit of infected vines. While there is evidence that the disease has been spread through grafted nursery stock, little is known about other methods by which the disease spreads within vineyards.

Grapevine fanleaf virus is thought to be the most serious viral disease of grapevines. Affected vines have leaves with malformed and abnormally gathered veins, giving the foliage a distinctive fanlike appearance. Infected vines decline over time, with severe loss of productivity and fruit quality. Primarily spread through contaminated nursery stock, movement through the vineyard is primarily via the feeding of dagger nematodes (*Xiphinema index*). While fanleaf degeneration has been found in Texas, ease of disease identification has resulted in immediate rouging of infected vines, which stopped further infection. There are also numerous other viruses, viruslike diseases, phytoplasmas, and other infectious agents that cause disease in vineyards around the world. When traveling abroad, suppress the temptation to return to your vineyard with "suitcase clones" from a vineyard you have visited. Plant quarantine exists for a reason and is our first line of defense against the next big disease problem.

Vineyard Floor Management

CONTROLLING competitive vegetation in the vineyard is the single most important component of establishing a vineyard. Weed scientists define a weed as a plant that is out of place. When weeds are taller than young vines, they compete for sunlight. Weeds are far more adapted to a given location and can beat young vines in the competition for any fertilizers applied, especially nitrogen. Competition for water, however, is the greatest threat uncontrolled weeds present, and the consequences can be disastrous. Managing competitive vegetation can be divided into separate tasks: managing weeds under the trellis row and managing them in the aisles, or row middles. Management options for accomplishing weed control vary by vine age, traditions of specific growing regions, and climate.

Before addressing weed growth, it is first important to understand how the roots of a grapevine grow and function. A healthy and dynamic root system is essential for the absorption of water and nutrients a grapevine needs. Roots grow only when there are ample amounts of water and oxygen in the root zone. Hardpan, plow pan, or waterlogged soils or anaerobic soil conditions limit the depth of rooting and the growth and productivity of grapevines. Cultural practices that compromise the root growth of a vine must be viewed in light of the limitations they present.

The Impact of Soil Cultivation

Since ancient times, tillage has been used to control competitive vegetation in vineyards, but in prior eras yield and quality expectations were lower and the consequences of tilling for soils and vines were not well understood. Some growers have made the statement that they are cultivating a vineyard in order to "make a vine grow deeper roots." This statement does not hold true, however; a vine will grow roots only as deep as

An example of superior vineyard floor management: a weed-free strip underneath the trellis and a closely mowed row center.

conditions exist to support root growth. Cutting off roots by tilling in the top few inches of soil is simply that—removal of an important part of a vine's functional root system. Although vines can produce roots as much as twenty-five feet deep, most of those roots are used for carbohydrate storage; the majority of the feeder roots are in the top six inches of soil. It's true that, in coarse soils, vines can root deeply and that feeder roots can extend as much as two feet below the surface, but tilling in the root system will still have negative consequences. There have been cases in which tillage of vineyard row centers caused a significant loss of feeder roots and significant crop loss as a result. It is believed such a reduction in yield is from both a lack of water at a critical time and an imbalance in normally occurring plant growth regulators. New root tips are an important source of cytokynins, so tilling of roots before bloom can result in a shortage of this plant hormone and severely reduce fruit set. Regular and intensive cultivation is also detrimental to the health of the vineyard floor. With cultivation, soil colloids are broken down, reducing soil porosity and air infiltration and causing a crust to form. In a tilled vineyard row center, equipment movement over cultivated soils will compact soil

Deep cultivation, or tillage, is still practiced in numerous vineyards.

◀ ▲ Cultivation severs important feeder roots in the top six inches of soil.

Fruit can fail to set because of a hormone imbalance brought about by deep cultivation just before bloom.

layers, especially in tractor wheel ruts, and thus constrict root growth. Perhaps the most significant impact of cultivation is that a vineyard becomes susceptible to erosion. Soils are a precious resource for growers, and scientists estimate that it takes approximately ten thousand years for one inch of topsoil to form. There have been cases in which vineyards that had been deeply cultivated then experienced sustained tropical rain events that resulted in losses of twelve to fifteen inches of soil. Obviously, those sites will never be the same.

One further implication of cultivating row centers is that after a rain event, it may be days before the vineyard dries out enough to accommodate tractor movement without causing serious ruts in the soil. Unfortunately, immediately after extended periods of rainfall growers may need to have quick access to the vineyard to apply fungicide sprays. While cultivation is still practiced, especially in arid regions where erosion is less of a threat, there are other ways to manage competing vegetation.

Herbicides

Among some sectors of the public, there is a stigma regarding the use of herbicides, as there is with many pesticides or other crop protection practices. Those sectors commonly associate herbicides with unsustainable farming practices. Perhaps the word *pesticide* should be considered in the same way we regard the word *drug*. If we consider the benefits of "drugs," are we talking about penicillin or heroin? The former has done worlds of good, with minimal social and environmental impact, while the other is generally perceived as having negative effects on society. Some pesticides have indeed caused problems for humans and the environment, while others are pretty close to being completely benign. If you use herbicides, or indeed any crop protection materials, take the time to

With cultivation, soil compaction frequently results in perennial ruts in the vineyard.

learn how they affect animal life and their fate in soils and what if any potential they have for negative impact. One can ardently defend the rights
of commercial growers to use pesticides, but that stance does not include
defending their misuse.

Managing Weeds under the Trellis

Traditionally, growers maintain a weed-free zone under the vineyard
trellis, and this zone typically is between twenty-four and forty inches in
width. Control of weeds under the trellis generally involves using pre-
emergence or post-emergence herbicides rather than mechanical cultiva-
tors. Pre-emergence herbicides can be applied alone or in mixtures to the
vineyard floor, where they act by inhibiting the germination and growth
of annual weeds. These materials and their application rates vary considerably by vine age, soil type, and soil chemistry. Some products may be
limited to nonbearing vineyards, while others are labeled for application
only on more mature vines.

In order to be effective, some pre-emergence materials must be incorporated into the soil by cultivation or rainfall. Some materials, such as
norflurazon (Solicam®), need at least three inches of rain to become fully
active. Consequently, while many pre-emergence herbicides are applied
in early spring, materials such as norflurazon can be applied in the fall to

increase the opportunity for incorporation by rainfall. Because materials and restrictions on their use change frequently, it is always wise to contact your local extension viticulturist for material options and specific product labels for up-to-date application guidelines.

Post-emergence herbicides are those that control weeds after they germinate or begin regrowth in the spring. Many post-emergence products, such as paraquat, glyphosate, and glufosinate, are nonselective and will burn back or kill any green tissue they contact. These materials are applied at various times during the year to kill or suppress annual and perennial weeds. Other spray materials, such as fluazifop-p-butyl (Fusilade®) and sethoxydim (Poast®), are selective grass killers. The selective grass killers can be sprayed in vineyards to control annual and perennial grasses without harming grapevines or any other broadleaf weeds. These grass killers offer growers an important tool, especially in managing perennial grasses such as Johnsongrass or Bermudagrass, but some materials are labeled for nonbearing vineyards only and others and must be applied long before harvest.

Potential or novice members of the grape-growing community should be aware of what is common knowledge among longtime growers: certain classes of hormonelike herbicides are extremely toxic to grapevines. Phenoxy herbicides such as 2,4-D, 2,4,5-T, dicamba, and others will kill or severely injure grapevines with minimal exposure. The most problematic of these materials is 2,4-D (2,4-Dichlorophenoxyacetic acid). Farmers and ranchers commonly use this material to control broadleaf weeds in grassy crops or to prevent weed growth by applying it prior to planting crops such as cotton. These materials can be extremely problematic because, once sprayed, 2,4-D readily volatilizes and can move great distances, even with only a gentle breeze. The use of these materials near vineyards is extremely controversial, and numerous commercial applicators refuse to use 2,4-D within a mile of a known vineyard location.

Understanding Herbicide Application

Fungicides and insecticides come with labeled rates, which means the specific rate or amount of material to be applied per acre of vineyard. However, recommendations for herbicide application are listed as the amount of product per *treated* acre. Thus, if a grower is treating a three-foot strip of floor under the trellis on rows spaced nine feet apart, only one-third of the vineyard floor receives treatment. In others words, for three acres of vineyard, there would be one treated acre.

Example of a grapevine expressing symptoms of 2,4-D exposure. Vines can be affected by spray applications of the herbicide from more than a mile away.

Application methods for herbicides can range from hand-held spray wands on backpack sprayers to booms mounted on four wheelers to stand-alone herbicide sprayers mounted on tractors. Application methods depend on how much area needs to be treated, the stability of a specific herbicide in suspension, and the precision at which the application needs to be made.

Applications of post-emergence materials such as glyphosate are commonly made from both mechanized and hand-held application equipment. With pre-emergence herbicides, accurate rates and placement are essential for proper weed control and vine safety, so those applications are best made from a tractor that can be set at a specific gear, sprayer pressure, and pump speed. In this case, an herbicide sprayer mounted either on the front or, more commonly, on the rear of the tractor is equipped with a pump driven by the tractor power take-off (PTO) and a spring-loaded boom that extends to the side of the tractor for application under the trellis. Spring-loaded booms are used so that, if the end of the boom inadvertently bumps a trunk or post, it swings away, then back into place. The following abbreviated checklist may help with accurately applying herbicides with this kind of equipment.

Vineyard herbicide sprayer behind a small tractor.

Easy Herbicide Sprayer Usage Checklist
1. Select the appropriate herbicide(s) and rate(s) for the expected weed spectrum and soil type present.
2. Thoroughly clean out spray tank, lines, and spray boom.
3. Partially fill spray tank with clean water, start pump, and set sprayer to desired pressure. Check spray nozzles for even pattern and uniform delivery. Measure width of spray band.
4. Measure off a test area, such as, for example, one hundred linear feet. With the tractor operating at desired ground speed and spray pump operating at desired pressure, determine how long it takes equipment to travel over the test area.
5. With the equipment stationary, run pump at the same speed and pressure and measure the volume of water that is delivered from each of the nozzles in the period of time it took to travel the test area. Determine the total volume from all nozzles. This total will give you a water volume-to-area ratio that will allow you to calculate the amount of water volume being applied per treated acre.

For example, assume the boom covers a 5-foot swath and it takes 20 seconds to travel the 100-foot test area. If there are four nozzles and each dispenses 15 ounces in 20 seconds, then 60 ounces of water are being applied over a 500-square-foot area. A simple ratio will allow you to calculate volume per treated acre. There are 43,560 square feet in an acre.

$$60/500 = x/43{,}560$$
$$60 \times 43{,}560 = 2{,}613{,}600$$
$$2{,}613{,}600/500 = 5{,}227.2 \text{ ounces per acre}$$
$$5227.2/128 \text{ (ounces per gallon)} = 40.8 \text{ gallons per acre}$$

If the amount of water is higher than desirable, a grower can use a higher gear or decrease operating pressure. If the amount of water is lower than desired, options for increasing the dispensing rate are to increase nozzle size, increase sprayer pressure, or use a lower gear.

With this knowledge, a grower will know how much material to put into the spray tank to correctly apply the right amount of material per acre. Remember that, while insecticides and fungicides are recommended for application at a given rate per acre of orchard or vineyard, herbicide recommendations are listed in amount of material per *treated* acre.

Row Center Management

With a clean strip underneath the trellis, vineyard row centers are perhaps best maintained through mowing and the selective use of post-emergence herbicides. The most efficient row center management plans begin with the use of a dormant season cover crop. Annual ryegrass, or perhaps oats, drilled or broadcast in the early fall will germinate with fall

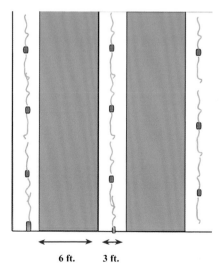

6 ft.   3 ft.

Effectively applying herbicides is a matter of matching spray volume to the amount of land treated. If vineyard rows are spaced 9 ft. apart and an herbicide is applied to a 3 ft. band underneath the trellis, one-third of the vineyard floor would be treated. A three-acre vineyard block would have one treated acre of land.

Example of a small-grain winter cover crop in a vineyard. Cover crops improve soil structure, support equipment mobility, and help control erosion.

rains and begin to grow as the weather begins to cool down. In wet winters, these cover crops may need to be mowed once or more to control their height, but the production and residue from this organic matter is a tremendous benefit to vineyard floors. Almost all of the soils in Texas and indeed across the South have very low organic matter content because, in the heat, existing microbes readily decompose organic residues.

The addition of organic matter improves soil structure, promotes a diverse soil microbe population, and can make some bound elements such as iron and zinc more available in high-pH soils. The easiest way to add organic matter to vineyard soil is to grow it there. Annual ryegrass planted in the fall is especially useful as a vineyard cover because, when it dies, the decaying ryegrass roots suppress the germination of other weeds. The ryegrass can be killed outright through the application of glyphosate right before bloom in the vineyard or over time, with the heat of summer. This process, called allelopathy, is a tremendous tool in the effort to suppress competing vegetation during the weeks following bloom, when berry cell division takes place. Because the flow of carbohydrates in a grapevine right before bloom is still from the root system up to the canopy, vines are at greatly reduced risk of injury as a result of incidental contact with systemic herbicides such as glyphosate. After bloom, it is a very different story, with vines being at a much greater risk of injury.

During spring frost events, however, tall-statured vineyard cover crops can impede heat accumulation by vineyard floors during the day and enhance radiational cooling in the mornings. When vines are susceptible to frost injury, tightly mowing or lightly cultivating the cover crop will help provide additional frost protection to the vines.

During the heat of the summer, most growers are content to manage the native vegetation that exists in their vineyard row centers by frequent mowing. In order to make vineyard floor vegetation unattractive to sharpshooters and other vectors of Pierce's disease, many growers keep the vineyard floor well manicured. The use of pre-emergence herbicides is not recommended for vineyard row centers. Having some vegetation in row centers is desirable because, with tractors and other equipment frequently passing up and down the vineyard rows, the grass acts as a cushion and helps reduce soil compaction.

CHAPTER 20

Grapevine Water Needs and Vineyard Irrigation

MANY GROWERS in eastern parts of the United States ponder the cost-effectiveness of irrigating a vineyard. Such locations may receive a minimum of forty inches and often more than fifty inches of precipitation a year. For all parts of Texas, however, even those receiving forty or more inches of rain annually, planning a commercial vineyard without the ability to apply supplemental water is at best unwise.

Average rainfall totals comprise rainfall rates from years that represent the extremes, both the very wet and the very dry. In areas with higher rainfall, the costs of drilling a well and the amount of energy needed to run irrigation pumps will be much lower than in more arid parts of the state, but at times it will still be necessary in order to achieve optimal vine growth and fruit maturation. Rainfall has the advantage of falling on the entire vineyard floor and supplying water to the entire root profile, which is necessary for root growth and the uptake of nutrients. Because irrigation supplies water to only a small portion of the root system of a vine, it is a poor substitute for natural rainfall but essential when it is necessary to supplement the amount of water provided to the vines by nature.

After a large rainfall event, vineyard floors are saturated with water. After water drains out of the large pore spaces, air can return to the soil. Under these conditions, the soil is considered to be at field capacity in terms of water, but the amount of water actually held will fluctuate considerably based on the texture and water-holding capacity of the soil, as well as rooting depth. Growers want air to quickly return to the soil because, in order for plants to absorb either water or nutrients, both water and air must be present in soils. Plants suffering from insufficient water and those suffering from excessive soil moisture both appear wilted and begin to develop nutritional deficiencies because of a lack of uptake ability due to root inactivity. Very coarse soils have the ability to drain quickly and thus allow air to return, but there is very limited lateral water movement through coarse soils and they have limited ability to retain large amounts of water for future vine use. Finer soils, such as clays, silts, and loams, drain more slowly, but lateral movement and water-holding capacity is better in those soils than in sandy or gravelly ones. Irrigation strategies for these very different soil types can be quite dissimilar.

As with most living organisms, plant vegetative material consists of 90 percent water. Water is needed in the root profile for nutrient uptake, utilization, and transportation within the vine. Under drip irrigation, in which only 10 to 15 percent of the root system has access to supplemental water, the vine can move water around the plant to compensate for reduced uptake from roots in drier regions of the soil. The vine itself cannot, however, redistribute water within the soil profile to allow for greater root function. Roots in dry soil may not die, but they will be minimally functional. During times of extended drought, it is common for vines to exhibit any number of nutritional deficiencies, especially nitrogen. In these cases, it is not really a shortage of nutrients but a shortage of functional, viable roots capable of taking up more nitrogen that is the issue. Nutritional needs can be addressed by injecting fertilizers into the drip system, but vineyards enduring periods of drought will not be as vigorous or productive as those receiving near-normal rainfall.

Water Application Methods

In many parts of the world, vineyards are irrigated through flooding or by overhead irrigation. For a number of reasons, neither of these techniques is suggested for use in Texas vineyards. Flood irrigation is a very inefficient use of water and cannot be counted on to supply a constant volume over a given vineyard block. Likewise, sprinkler irrigation is also an inefficient use of water. In addition, the wetting of the canopy along with the soil typically results in greatly increased fungal disease pressure. Totally saturating the soil through either of these methods also means that putting equipment into the vineyard would cause severe rutting and compaction of the vineyard floor. In the Texas climate, growers cannot afford to stay off the vineyard floor for several days. They need to be able to get into the field at a moment's notice to apply timely sprays because a delay of a day or more can be disastrous.

Drip irrigation has been the standard for supplemental applications of water in vineyards for decades. A drip system allows for water application at precise points throughout the vineyard, with minimal water loss due to evaporation. The limitation of a drip system is simply the percentage of roots that have access to water. Microsprinklers, which discharge water over a four- to six-foot area under the vine, can be employed to overcome this limitation, but evaporation loss is higher and few growers actually have the luxury of applying that much water. Drip irrigation usually flows from a well, with the water being drawn by a pump through a filter, then discharged into main lines that have risers at each row. Polyethylene pipe, usually one-half to three-quarters of an inch in diameter, runs down each row to deliver water under each vine.

In newly established vineyards (first and second growing seasons), one emitter per vine is sufficient, and each emitter is typically placed twelve to fifteen inches away from a trunk. It is important that the emitter not be placed directly at the trunk because not only is that not where the feeder roots are but saturation of the crown can lead to increased disease pressure. In the third growing season, the lateral lines are moved so that the old emitter is twenty-four to thirty inches away from the trunk and a second emitter is then placed on the other side of the vine.

Growers can choose from among a wide array of drip emitters, with flow rates that range from one-half gallon to two gallons per hour. Drip line with emitters already installed within the pipe is also available. With internal emitters spaced closely within the line, these pipes can be used to wet a solid strip of soil underneath the trellis. While this type of drip line may waste water in young vineyards, it can be a distinct advantage in mature vineyards.

In some parts of the state, particularly the High Plains, burying irrigation pipe twelve to fifteen inches below the surface is a common practice. Where soil-burrowing voles and gophers are not present, this practice further reduces loss from evaporation and will reduce the shrink/swell movement of irrigation laterals caused by seasonal changes in temperature. It does, however, have the disadvantage that breaks or clogging of irrigation lines will not be noticed until vines have a visible response to too much or not enough water.

It is highly recommended that new growers seek the advice and services of a competent irrigation specialist who can accurately and economically design a system to meet individual needs. While you may choose to do the majority of the installation work yourself, a professional should be consulted for irrigation system design and material selection. Most large-scale irrigation dealers have irrigation engineers on staff. If you purchase your materials from that dealer, the services of the engineer are typically available at little or no cost to the grower.

◀ Irrigation main lines are typically buried underground from the well to a riser at each row.

▼ Polyethylene pipe, from one-half to three-quarters of an inch in diameter, is typically hung from a low trellis wire to deliver water to each vine.

In first- and second-leaf vines, usually one emitter is used per vine, and it is placed twelve to fifteen inches away from the trunk to deliver water to the zone of new root growth.

Water Sources

Most vineyards that are irrigated utilize wells to provide water for the irrigation system. Surface water sources can be used, but there are a number of logistical and regulatory drawbacks involved. While some growers have limited access to river water, the days of being able to utilize that source are probably numbered, and having a site close enough to a river for the grower to economically pump water may not be advisable simply because of frost risk. Ponds or lakes have been used as sources of irrigation water, but they have limitations as well. Surface water from ponds typically has a lot of organic and inorganic material in suspension and must be highly filtered to avoid clogging emitters or microsprinkler heads. It also takes a very large lake or pond to supply the needs of a modest vineyard. For example, if a grower wants to apply 10 gallons of water per vine per week in a 10-acre vineyard and vines are spaced 9 feet by 8 feet, that vineyard would require 60,500 gallons of water per week. A pond 100 feet in diameter and 8 feet deep would hold 62,800 cubic feet, or 470,000 gallons, of water, which would last just under eight weeks. While ponds allow for quick access to large amounts of water, they are better utilized as sources of water for a sprinkler system designed for frost control rather than day-to-day irrigation.

The well capacity of a given site should be enough to support a mature vineyard with a full canopy and crop during the heat of the worst summer imaginable. A well capacity of five gallons per minute per acre is a conservative benchmark for vineyard establishment. A well with a fifty-gallon-per-minute capacity can pump seventy-two thousand gallons a day, enough to support vine growth and crop maturation on ten acres during severe weather conditions.

Wells should also be tested for water quality. Water must have an SAR (sodium absorption ratio) below 7, preferably below 4. Grapevines are sensitive to mildly elevated sodium, chlorine, and boron levels in water. Chlorine content should be below 15 milliequivalent/liter, boron should be below 0.5 parts per million (ppm), and, while the SAR is more indicative of the threat of sodium, total ppm is ideally below 1,000 ppm. While neighboring wells may give some insight into local water quality, each well is unique and should be tested separately.

Determining Water Needs

There are numerous ways of determining water needs within a vineyard, and some are far more sophisticated than others. This text is intended to be neither a comprehensive primer on which method is superior nor a guide to walk you through the mathematical calculations of each technique. It is important to know that the mechanical and mathematical methods of estimating water loss and water availability are just that—estimates. The appropriate amount of water in a vineyard soil profile is dependent on variety, rootstock, vineyard age, the time of year, crop load, current and projected weather conditions, and myriad other factors. It is important to learn what these tools can do for you, but it is equally important for a grower to learn how to look at the vines and discern how water availability is affecting growth.

Gypsum blocks have been in use in vineyards since the 1940s and are still one of the most economical, reliable ways of measuring available soil moisture. Blocks are permanently buried in the ground at different depths, perhaps twelve and twenty-four inches, with wire leads coming back above the surface of the vineyard floor. These wires are connected to a meter that measures the resistance needed to pull moisture out of the soil profile into the gypsum block. Numerous measurement stations are established within and between vineyard blocks, depending on soil texture and depth. Over time, growers can use these measurements to begin to discern at what point vines will benefit from supplemental irrigation.

Using pan evaporation measurements is a way of estimating the overall water loss from a site. This method allows growers to know approximately how much water needs to be applied on a daily or weekly basis. Evapotranspiration is the loss of water from a site based on the transpiration of a particular crop plus the amount of water lost in the environment due to evaporation. Weather stations around the state can provide real-time pan evaporation figures, and, when these figures are combined with the specific coefficient for grapevines, the grower can calculate an estimate of how much water per day needs to be applied to maintain a particular level of water availability in the vineyard.

Perhaps the single most reliable way of determining grapevine water status is to simply look at the vines. Accurately judging water needs based on simple observation is a skill acquired over time, but lack of sufficient water can affect vine growth at any time of the year. Very dry conditions can seriously delay or protract budbreak, leading to problems with uniform fruit maturity at harvest. During the growing season, the first manifestation of drought in the vines is usually in the appearance of leaves. Photosynthesis is limited when temperatures exceed 95 degrees Fahrenheit, but with ample water, leaves can transpire water through the stomata on their lower side, thus modifying leaf tissue temperature through evaporative cooling. While it is common under very high temperatures for leaves to become somewhat wilted, a lack of water will slow recovery time once temperatures cool.

Shoot tips and tendrils are perhaps the greatest indicator of vine water status. If shoot tips are actively growing during the spring and early summer and tendril length is at least two inches, the vines likely have adequate water. If, during this period of grand growth, shoot tips begin to become small and leathery, internode length starts to become markedly smaller, and distal tendril length falls below two inches, insufficient water is indicated. When shoot tips abscise and tendrils begin to become necrotic, water availability is dangerously low. In the Texas climate, water is a tremendous tool for controlling vine vigor, but, like most things, it's a balancing act. Too much and the vines become excessively vegetative, too little and growth and fruit quality are negatively affected. If there is high vegetative growth going into véraison, the foliage, not the fruit, becomes the primary photosynthetic sink. By restricting water to a degree, vines place a larger amount of their photosynthates into the fruit rather than new growth.

Of particular note is the repeated advice of some "experts" on water availability status as harvest approaches. All too commonly growers are given the advice of shutting off the water after véraison to hasten maturity. In Texas, temperatures commonly exceed 100 degrees Fahrenheit during that period. Sugar accumulation in berries is a direct consequence of photosynthesis. With inadequate water, vines lose the ability to cool their leaves, leading to reduced photosynthesis and delayed fruit maturity. There have been times that the lack of available water has led to vineyard defoliation, which had profoundly negative effects on fruit quality and subsequent vine winter hardiness. Growers need to avoid the extremes of moisture status and try to strike the balance of optimal vine maintenance and fruit maturity.

Insects and Mite Pests

Phylloxera

RAPE PHYLLOXERA (*Daktulosphaira vitifoliae* [Fitch]), a small, aphidlike insect pest, is native to Texas and most of North America, where it has evolved alongside native North American grapevine species for tens of thousands of years. European wine grape varieties derived from *Vitis vinifera* evolved in areas without this insect pest and consequently exhibit varying degrees of susceptibility.

There are two forms of phylloxera: the foliar form, which feeds and reproduces on grape leaves, and the root form, which kills new feeder roots and, on very susceptible varieties, can kill older anchorage roots. The foliar form can be seen on both cultivated and native grape species. In commercial vineyards the foliar form is typically controlled by applying two insecticide sprays, at approximately ten-day intervals, when galling is first seen. The root form is far more damaging and must be controlled by the use of resistant rootstocks.

The introduction of the New World pest phylloxera into Europe in the late 1850s devastated the European wine industry, and the insect and vineyard devastation spread through North Africa and into South America. Introduction of hybrid phylloxera-resistant rootstocks from North America ultimately provided the control necessary for the revitalization of grape growing in Europe, and these rootstocks are still employed in vineyards around the world today. While purchasing own-rooted vines may seem like a way to save money when establishing a vineyard, new growers must remember that phylloxera is native to Texas and that once a small resident population of this insect finds an own-rooted vineyard, it is just a matter of time before widespread infestation takes its toll on the vines.

Early stages of gall formation from a phylloxera infestation.

Sharpshooters

There are more than thirty species of insects known as sharpshooter leafhoppers that are native to different parts of Texas. The two major tribes of the subfamily Cicadellinae are the larger, more distant-flying Proconiini (of which the glassy-winged sharpshooter is a member), and the smaller, vineyard-edge feeding Cicadellini.

All of these xylem-feeding insects represent very competent vectors of Pierce's disease, and they all exist in every part of the state. Other than the risk they pose through the transmission of PD, the feeding of these insects poses no risk to the health and productivity of grapevines. To native grapevine species and to PD resistant grape varieties, these insects are of little consequence. However, to growers of susceptible varieties, thorough knowledge of these insects and their abundance, preferred habitats, and seasonality is critical information in the management of Pierce's disease. Detailed information on these insects and their control can be found in the 2012 publication *Pierce's Disease Overview & Management Guide: A Resource for Grape Growers in Texas and Other Eastern U.S.*

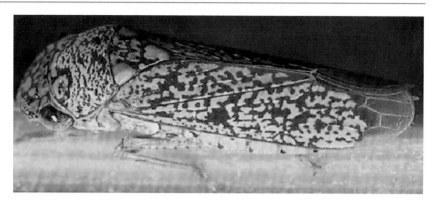

Oncometopia orbona, one of many competent species of xylem-feeding sharpshooters capable of transmitting Pierce's disease.

Growing Regions, which can easily be found on the web through common search engines.

Grape Berry Moth

While undoubtedly the grape berry moth (also known as GBM, or *Paralobesia viteana* [Clemens]) has in the past existed on wild grapevines throughout the state, the intensified cultivation of commercial vineyards has resulted in this insect becoming an important pest of grapevines across the state. GBM overwinters as a pupa and in the spring emerges as an adult, mates, and lays its first generation of eggs on flower clusters or young fruit. Pupae are commonly blown to wooded edges of vineyards, particularly the southern and eastern edge, so there is usually more injury to fruit in the first five or six rows adjacent to these wooded edges. After eggs hatch, the emerging larvae feed on flowers or fruit, which can result in significant economic loss. Second and subsequent generations of this pest do similar damage, but because berries are larger, each larva usually infests only one or two berries. In addition to causing direct crop loss, GBM feeding on berries from midseason to ripening time provides an entry site for numerous fungal and bacterial decay organisms, which can significantly reduce yield or seriously compromise fruit quality.

Control efforts typically begin with insecticide applications seven to ten days after full bloom, but exact timing of these and subsequent sprays can be fine-tuned with the use of pheromone traps that can help monitor male moth flight. New GBM prediction models have been developed in New York, Pennsylvania, and Michigan that have successfully pre-

Grape berry moth feeding in developing grape flower cluster.

Second-generation grape berry moth feeding on developing berries.

Pheromone trap being deployed in a grape vineyard to help monitor male moth emergence, which signals the best time to target sprays for control of hatching eggs.

dicted emerging generations based on growing degree models compared to bloom dates in native grapevines. Numerous publications from Texas and other eastern US grape-growing regions should be consulted for additional GBM-specific management options.

Leafhoppers

In the same family as sharpshooters, leafhoppers (*Erythroneura* spp.) are insects that belong to the family Cicadellidae, which comprises several species of phloem-feeding, short-flying insects. These pests feed primarily on foliage by extracting juices from leaves, which subsequently kills localized areas of cells and thus reduces the photosynthetic capacity of the grapevine canopy. Like other diseases or deficiencies that deprive the vine of chlorophyll, the amount of damage to the vine is proportional to the amount of photosynthetic area affected.

While small numbers of these insects can be tolerated, thresholds for applying control measures vary by variety, crop load, and region of the country. While leafhoppers can be found in every grape-growing

Leaf bronzing caused by phloem-feeding leafhoppers attacking the underside of leaves. Leafhoppers suck juices from leaf cells, thereby reducing the photosynthetic capacity of the leaf.

region of the state, West Texas and the High Plains appear to have the highest populations of this pest. Leafhoppers overwinter as unmated adults in leaf litter, and in the spring these adults emerge from quiescence and begin feeding on grapevine suckers, basal leaf tissue, or other vineyard vegetation. As temperatures warm, adults mate and oviposition takes place as eggs are laid on leaf tissue. Nymphs emerge in seven to fourteen days, depending on temperature.

Nymphs go through five flightless instar stages before becoming adults, but both nymphs and adults feed on green grapevine tissue. In many cases, insecticide applications timed for first-generation GBM can also greatly assist in the management of leafhoppers, but growers should guard against late-season infestations of leafhoppers, which can cause premature defoliation and lead to a reduction in vine cold hardiness and vigor.

Climbing Cutworms

Climbing cutworms are the larval stage of a number of species of moths that exhibit nighttime feeding. The different species vary slightly in their appearance but are generally brown or gray, with stripes running lengthwise down their smooth-skinned bodies. Cutworms are usually not seen during the day and remain in weed stubble or organic mulch until evening, when they climb up vines and feed on small, developing shoots. Grapevines remain susceptible to feeding injury until shoots reach approximately three inches in length, but the period of susceptibility is commonly extended when cool spring temperatures slow early-season shoot growth.

Damage from climbing cutworms appears as hollowed out buds or shoots and is commonly overlooked until economic injury has occurred. Because each primary shoot eaten represents an average loss of two clus-

Climbing cutworm making a rare daytime appearance. Cutworms hide under leaf litter and typically come out at night to feed.

Young shoot eaten by climbing cutworm. Each bud eaten typically results in a two-cluster yield loss for a vine.

ters per vine, the economic threshold for control measures is reached when the first couple of vines have been spotted with feeding injury. For every noticeable feeding foray, there are certainly ten more you have not yet seen. A weed-free strip under the trellis helps reduce the availability of cover for cutworms during the day and can lessen feeding pressure. Conversely, while applications of mulch under the trellis have numerous viticultural advantages, they offer additional daytime refuge for cutworms.

Green June Beetles

The brightly colored true beetles known as green June beetles (*Cotinus nitida* [Linnaeus]) can be highly problematic for ripe or near-ripe fruit. In addition to causing direct losses of fruit from feeding, these insects cause berry injury that, especially when combined with marginal precipitation, can lead to widespread fruit infection from fungal and bacterial rot organisms.

Green June beetles overwinter as grubs as much as a foot underground and in the early spring slowly make their way to the surface of the soil. At this stage they are feeding on decaying organic matter and occasionally on grape roots. The grubs pupate in late spring, and the adults emerge in early summer to feed on many types of foliage,

Metallic green June beetles feeding in a fruit cluster.

but they are especially drawn to various types of ripening fruit. Once feeding begins, apparently a feeding pheromone is released, and large numbers of beetles appear to join the original perpetrators.

In addition to causing direct fruit injury, green June beetles, even in small numbers, can affect flavor in fruit, which leads to defective wines. Because of the threat of secondary rot organisms, control measures should be considered when feeding is first noticed. Selecting spray materials that can be used until relatively close to harvest is usually necessary to reduce fruit losses immediately before harvest .

Grape Cane Borer

During dormant pruning, growers often notice canes that have holes slightly below the leaf scar at the node, approximately midshoot on a dormant cane. These are most commonly the exit holes of grape cane borers that have infested green shoots during the previous growing season. They are generally of little consequence in cordon pruned vines. Because injury occurs only occasionally in vineyards, these infestations are not considered economically important. However, in some vineyard locations, especially along wooded or brushy vineyard edges, infestation levels can be high and cane breakage may be problematic when a grower is trying to establish new or replacement cordons. When and where needed, control

Exit hole left by an emerging grape cane borer.

measures are typically timed at five- to ten-inch shoot growth to control larvae as they hatch from eggs on young developing shoots.

Leaf Folders and Skeletonizers

A number of insects from the order Lepidoptera can and do inflict economic injury to grapevines by feeding on or damaging foliage. These larval pests include grape leaf roller, grape leaf folder, grapeleaf skeletonizer, and western grape skeletonizer. Grape leaf roller and grape leaf folder use silklike excrement to encase themselves in leaf tissue, which provides protection from predation and in many cases from contact insecticides.

The two grape skeletonizer species that are found in Texas are members of the Zygaenidae family of moths, which have the capability of producing hydrogen cyanide, a potent deterrent of birds and other predators. Because they commonly appear in large numbers, these insects all have the capability of inflicting considerable damage to foliage in a short period of time. Control measures should begin when widespread feeding becomes apparent. Since the damaging forms of these pests are all larval stages of moths, the organic insecticide *Bacillus thuringiensis* (Bt) is effective against them if applied in a timely manner. Bt is short-lived in the environment, and repeat applications may be required for the desired level of control.

Grape leaf folder feeding on terminal tissue. Leaf folders exhibit strong varietal preference in feeding.

Minor Insect Pests

There are many, many insects that can be seen in and around vineyards, and even though isolated cases may be found in which they are actively feeding on foliage, many of them are of little or no economic importance. Plume moth infestations, while quite noticeable, usually involve the occasional folding and ingestion of one or two new leaves

Grapevine epimenis, a casual feeder on grape tissue, rarely warrants control.

on developing shoots, but plume moth has never been shown to cause reductions in growth, vine yield, or fruit quality. Other insects found in Texas in that group are grape cane gallmaker, grape cane girdler, grape tumid gallmaker, and grapevine epimenis, along with many others.

While insecticides, whether organic or conventional, may at times be needed to control economically damaging insects, in many cases insecticide applications are not warranted for rare or isolated pests and new growers need to learn to not overreact. Still, there are numerous pests that can have a negative economic impact on grapevines found in other parts of the country, and they have apparently not yet found their way to Texas. When encountering new potential insect problems, growers should make every effort to capture or photograph the insect and record or photograph the suspected damage before discussing the situation with your extension specialist or vineyard consultant.

Mites

Mites are not insects but members of the Arachnida class (which includes spiders), and mite damage on grapevines is rather rare in Texas, although other crops are susceptible. The two species of mites most commonly found in Texas are the two-spotted spider mite and European red mite. Spider mites colonize the underside of leaf tissue and congregate along leaf veins. Mite injury to foliage appears as a subtle stippling along the leaf vein, in some cases resembling leafhopper injury.

Mite outbreaks may be prompted by hot, dry, dusty conditions, but the most typical cause of flares is overapplication of carbamate, pyrethroid, or nicotenoid insecticides to the foliage. These insecticides are broad-spectrum products that do not kill mites but instead kill mite predators, thus allowing for unabated mite reproduction and then colonization of green tissue.

There many species of mites that do commonly reside on the underside of grape leaf tissue in Texas, but most species do not cause harm to the plant. Mites have a short life cycle, can reproduce very quickly, and cause severe economic injury to a vineyard canopy in a short period of time. If control measures are necessary, depending on the miticide product used, two applications seven to ten days apart may be needed to provide sufficient control to avoid economic damage. If mite infestation is suspected, support can be obtained through the cooperative extension system for identification of the pest and recommendations on control measures.

CHAPTER 22

Vertebrate Pest Control

THERE ARE a number of God's creatures that love your vines every bit as much as you do. If you plan on growing grapes where there is even a small deer population, erecting a game fence is a necessity. Grapevines are among deer's most favored forages. They will eat every leaf that grows, and a small deer population will turn into a large one. Chemical deer repellents are available and might deter some deer in some years, but in dry years they are basically ineffective. Sound and motion scare devices may work for a while, but deer will no longer be intimidated once they become used to the device.

Game fencing will be required in much of the state to prevent damage from deer, hogs, raccoons, and many other animals that enjoy eating grape foliage and fruit.

There is no such thing as a deer-proof fence, but some fences do a good job of keeping deer out of your vineyard. Deer fences need to be at least eight feet in height, and most are constructed of ten-foot t-posts. Use either twenty strands of barbed wire or two sections of four-foot goat wire, tied one above the other on the posts. Barbed wire at the ground and again at four to six inches above the top goat wire section helps deter diggers and jumpers.

Likewise, horses, cattle, sheep, goats, and llamas will all eat grapevines, so the idea of your favorite mammal living in harmony with your vineyard is a fantasy. If they don't eat the vines, they will tear down or break something. Rabbits can be a severe problem in newly planted vineyards, and a smaller mesh wire, such as eighteen-inch chicken wire, attached to the bottom of the deer fence may be needed to keep rabbits out. If there are feral hogs in your area, be prepared to have your defenses challenged on a regular basis and occasionally broken through. Where they exist, feral hogs pose the most devastating, destructive hazard to both crop and vine health that a Texas grape grower may face. The only larger mammalian foe is the occasional bear that challenges plantings in the Trans-Pecos region.

Any number of animals will come into your vineyard right before harvest and beat you to the punch. Raccoons, opossums, and squirrels can climb or slip through the deer fence and proceed to consume or destroy an enormous amount of fruit in a very short period of time. Chicken wire will in most cases deter squirrels, but raccoons and opossums usually stop only for a pair of electrified wires placed outside of the deer fence. These systems can be very effective at deterring these predators, but because grasses and broadleaved weeds can ground out the electrified wires, a concerted effort to control vegetation under and around the wires is needed to keep these systems functional.

Perhaps the most problematic vertebrate pests are birds. Seed eaters such as doves commonly nest in vineyard canopies, but they pose no threat because they don't eat fruit. There are a number of bird species, however, that absolutely love to eat ripening grapes. Starlings and, unfortunately, mockingbirds (the state bird and a protected species) are perhaps the most problematic. Vineyards in proximity to tree lines or power lines will most likely experience the highest bird pressure, whereas growers on the High Plains face relatively little fruit damage due to birds. Sound devices such as carbide cannons (not popular with neighbors!)

or bird distress calls have had some measured success in mitigating bird pressure. As with deer, bird scare alarms work for a while, but unless the distress calls are changed frequently, bird populations become used to the racket and eat the fruit anyway. Reflective tape, plastic hawk and owl replicas, and raptor kites all may be of some help under very low bird pressure, but they commonly only delay the inevitable.

The only real solution to bird control is to net the vineyard. Bird netting typically consists of a nylon or polypropylene mesh draping that is applied over the canopy and tied below the cordon wire to keep birds away from the fruiting zone. Traditional netting is from fourteen to seventeen feet in width, with mesh sizes from one-quarter to one inch in size, and available in rolls of varying lengths.

Netting is usually applied over the rows relatively soon after véraison, when the fruit begins to become attractive to the birds. Ideally, the vegetative growth of grapevines has slowed or stopped prior to application because shoot tips growing through this netting become problematic when the netting needs to be removed immediately preceding harvest. Bird netting can be deployed by hand with relative ease, but mechanical aids have been developed that are mounted on a tractor's front end forks and hold the netting roll above the row. Some growers take the netting

In areas with bird pressure, bird netting is the most reliable way of preventing serious economic losses.

Mechanical devices are available that assist in the deployment of bird netting.

off the vines and tie it to the wire holding the irrigation laterals until the following season. Growers seeking to minimize the problem of netting becoming brittle due to ultraviolet light exposure may remove netting each year and store it in plastic bags. Netting can be a considerable expense to buy, deploy, and store, but without protection from birds, entire crops can be lost.

CHAPTER 23

Vineyard Equipment and Infrastructure Needs

I N ORDER to have a successful vineyard, one needs more equipment than a pair of pruning shears and a wire stretcher, but growers just starting out will find that the economies of scale work against them. While corners can be cut in the first couple of years, growers who are serious about growing a grape crop of high commercial quality will need to invest in tractors and sprayer equipment capable of not only caring for the vineyard but also, and in particular, protecting the crop.

Tractors

Tractors are the backbone of any farming operation. It is not uncommon for new growers to get smaller, used tractors for initial vineyard establishment and then trade the old tractor in for a new(er), more powerful one once they have need of an airblast sprayer. A tractor does need to be dependable. Many new growers establish vineyards on a parttime basis, so their time is limited, but when they are at the vineyard site, time is very valuable. It is incredibly frustrating to arrive on site, then be thwarted in your efforts because the tractor won't start or has mechanical limitations. Unlike the common wisdom for machines used to grow agronomic crops, bigger is certainly not better when it comes to machinery for vineyards. Smaller, more efficient equipment is preferable for navigating the sometimes tight confines of a vineyard in full production.

Tractors are generally rated by horsepower and by PTO (power takeoff) horsepower. The need for pulling or plowing in a vineyard is minimal, so the ability of a tractor to run a sprayer pump, which requires PTO horsepower, is the more critical factor. Tractors need a minimum of 35 to 40 PTO horsepower to operate the sprayer and fan of a modest airblast sprayer. Although 50 horsepower is better, you may not need that much power unless your sprayer requires it. Many equipment manufacturers make specialized vineyard and orchard tractors that have a narrow, low profile designed to move through tightly spaced rows. These tractors are more expensive than the run-of-the-mill tractors commonly used for farm and ranch work, but growers generally deem them worth the money.

Cabs on tractors are also becoming more common. Although introduced primarily to protect the driver from chemical applications, cabs also provide air-conditioning and protection from sunlight and wayward canes in the vineyard—all good reasons to consider a cab on your tractor. Again, they are a hefty expense, but think what that cab will have meant to you in ten years.

Most common farm tractors are equipped with narrow, deeply grooved tires to provide traction and keep the tractor mobile in deep mud and on steep terrain. Vineyard and orchard tractors are commonly equipped with aircraft or "turf" tires that float over a vineyard row rather than dig through the row. These tires are far superior for all vineyard equipment because they significantly reduce compaction of the vineyard floor.

Mowers

Every grower will need equipment to mow vegetation around the vineyard and in the row centers. The most common mower is known as

A reliable tractor with sufficient horsepower for the jobs at hand is essential in a commercial vineyard operation.

Brush choppers are commonly used to keep vegetation short in row centers and alleyways.

a brush hog, brush chopper, or shredder. It consists of one or two blades turned within a confined box pulled behind the tractor and powered by the PTO shaft. Generally, the PTO needs of such equipment are minimal, and 17- to 25-horsepower tractors can do a sufficient job of powering this mower.

These mowers come in various widths, and the dimension you need should be carefully considered before purchase. While a wider mower is more difficult to maneuver through the vineyard, significant time will be saved over the life of the vineyard if the row center can be mowed with a single pass. If rows are spaced nine feet apart and there is a two-foot weed-free band under the trellis, a six-foot mower will not be wide enough to cut the row center with one pass. In this situation, the grower must either expand the weed-free band under the trellis or consider buying a seven-foot mower.

Other types of mowers, such as flail mowers, are also commonly employed. Flail mowers consist of many L-shaped blades that, using centrifugal force, chop up vegetation as the mower passes over it. Flail mowers are generally considered superior for chopping up pruning brush and give the vineyard floor a more manicured look, but they require more maintenance than simple shredders.

Three-point hitch mounted airblast sprayers that do not require higher horsepower tractors to operate are available for smaller vineyard operations.

Airblast Sprayers

While fungicide applications on first- and second-leaf vineyards are commonly made with simple hand-held equipment of some type, by the time a vineyard is about to bear its first crop fungicide application must be taken to the next level. There are numerous types of air-assisted sprayers, but the most common variety uses a pump that is powered by the power take-off shaft of the tractor. The pump pressurizes spray material in the tank to a specific level, and the material is then moved to a series of nozzles arranged on one or two banks at the rear of the sprayer. Each bank of nozzles consists of five to seven ports, which have whirl plates and nozzles that break the spray material into small droplets. These ports can be moved vertically to create a spray pattern that will put the material where it is needed. The PTO shaft also turns a fan that amplifies the power at which the spray material is delivered to the canopy of a vineyard row.

Airblast sprayers with tank capacities from 50 to 150 gallons are usually mounted on the three-point hitch of the tractor, while larger sprayers are usually mounted on trailers pulled behind the tractor. It is also important for airblast sprayers to have some ability to keep the chemical agents suspended in the water. Mechanical paddle agitators within the tank will accomplish this task, or it can be done with return flow, in which spray material is drawn into the system by the pump, then forcibly returned to the tank, creating a mixing action.

It is important for growers to realize that the goal is to thoroughly drive the spray material into the canopy, not simply to spray the material into the air and hope it lands where it is needed. It is far better to have banks on both sides of the sprayer because spraying can then be finished in half the time, whereas a single-bank sprayer must travel each row twice, once in each direction, in order to spray the vineyard thoroughly. Sprayers with two banks of nozzles also allow a grower to more safely apply spray materials.

For materials to be accurately applied while minimizing off-site spray drift, spraying needs to be done when winds are low, commonly at dusk, during the night, or very early in the morning. Even during these times, there will be a slight breeze during an application, and a sprayer with two sets of nozzles has a distinct advantage in this situation. In order to minimize exposure of the vineyard worker to the spray, the equipment operator can spray into the mild breeze and create on each side a slight vortex that whirls through the canopy. When spraying into the wind, the operator should stop the tractor at the end of

Wind shear airblast sprayers are more expensive than blade fan sprayers, but many growers consider them more effective in applying spray coverage.

the row, then stop the sprayer and deadhead (i.e., drive in the leeward direction without spraying and start again on another row, spraying once again into the wind). Application coverage can be improved and operator exposure can be minimized with this technique.

There are many publications on spray application safety, and it is strongly recommended that anyone getting into this business be thoroughly educated in these practices. The person at greatest risk of pesticide exposure is most certainly the applicator, and the time of greatest risk is during mixing and handling of material going into the tank. Understanding and following proper handling procedures is absolutely essential in minimizing the risks of handling pesticides.

Herbicide Sprayers

Many growers, especially those with more than a couple of acres, will have a tractor-mounted herbicide sprayer. With traditional herbicide sprayers, tanks are usually mounted on the three-point hitch behind the tractor and powered by a roller or piston pump mounted on the tractor PTO. Spray material can be applied with a shielded boom mounted near the vineyard floor, through an open boom mounted on one side of the tractor, or by a hand wand.

These sprayers can be accurately calibrated and are superior in applying pre-emergence herbicides in bands underneath the trellis. Booms consist of a series of flat fan nozzles that vary by how much material they apply and by the angle of the triangle that delivers the spray material. Post-emergence herbicides are also commonly applied through these systems. The recommended amount of water carrier per acre varies greatly between these different materials, so spray tip selection can be utilized to greatly increase or decrease the amount of water the boom delivers.

While this book does not propose to endorse specific equipment, it is worth mentioning that sprayers such as Bubco® have greatly improved the ability of growers to apply herbicides with nominal or no threat to vines. These products utilize controlled-droplet application sprayers with spring-loaded covers that allow for herbicide application with virtually no off-target drift. This technology means that post-emergence herbicides such as glyphosate can be applied much later in the season than normally thought safe because they greatly reduce the danger of vine injury as a result of incidental contact. These sprayers are not inexpensive but are very effective tools that will allow for herbicide cost savings as compared to the use of other materials that are much more expensive.

No-Till Drill

While many may not think of the no-till drill as essential equipment, without it there is virtually no way to plant winter cover crops without tillage of vineyard rows. As was outlined in the vineyard floor management chapter, tillage has many negative effects, perhaps the greatest being the loss of important grapevine feeder roots. Vineyard floor covers must be managed during the growing season so that grass and weeds do not compete with the vines for water, but as temperatures cool and vines go dormant, the water requirements of grapevines are greatly reduced. At that point, winter cover crops such as annual ryegrass, elbon rye, or even oats are planted to generate the organic matter that Texas soils lack and thus to improve soil structure and microbial diversity.

These grasses are typically planted in early fall, but for these crops to efficiently germinate, there must be excellent seed/soil contact. Seeds cannot simply be broadcast across the vineyard floor; they must be placed into the soil. No-till drills utilize wheels that cut through the thatch of organic matter. Accurately metered seed-boxes with tubes extend into the ground and insert the seeds, and wheels then replace the soil and thatch over the top of the seed. This equipment plants cover crops in rows four to six inches apart and can greatly improve germination efficiency in a cover crop. Because this equipment is not used much during the growing

A no-till seeder can allow for the planting of winter cover crops without the harmful effects of tillage.

season, it is very feasible for a group of growers to buy this equipment together and use it cooperatively.

Other Equipment

A number of other items can be useful in a commercial or even a small vineyard operation. Utility vehicles with a roof and a bed can be invaluable in taking supplies or people to and from the vineyard. They are also quite useful when doing routine tasks that require traveling across the whole vineyard, such as inspecting irrigation lines in the spring or summer.

Another handy tool is a box blade, which can be used when trying to eliminate ruts in roads or to fill low spots in vineyards to keep them from becoming deeper or larger. In some more northern areas, some type of grape hoe is used to move soil over the vine graft union (in a process called "hilling") to protect it during the winter months and then to remove the soil in the spring. While necessary in some areas, grape hoes are the cause of much mechanical injury to the crowns of grapevines, which is never a good thing. Shoot positioners, leaf pullers, brush rakes—the list goes on. There is a never-ending list of tools that can be of help in a vineyard, but it's a list that at some point is usually cut short by financial considerations.

Various types of utility vehicles are now commonly used to apply contact herbicides.

Sheds

Every vineyard operation needs a place to keep equipment out of the weather. Tractors are the most important thing to keep covered, but the life of any piece of equipment will be increased if it is protected from the elements. Work benches for equipment repair, shelves to hold supplies, tools, and perhaps raingear are all important. Having a shed or barn where you can get out of the weather every once in a while will be important for your long-term health and the longevity of any vineyard workers you might find. *A separate, locked storage shed, properly signed and protected from the weather, is needed for pesticide storage.* Keep your tools well maintained and your workers warm and dry, and your vineyard operation will be much more productive.

Weather Station

It is important for growers to monitor weather conditions and to be able to compare detailed notes on how weather affects the vineyard from year to year. For a reasonable price growers can now purchase weather stations that can record temperature, relative humidity, and rainfall, measure leaf wetness within canopies, and predict disease outbreaks in the

A barn to protect equipment and to house a workshop is a big plus in maintaining and repairing vineyard equipment.

vineyard. Rather than depending on hand written notes, growers can use these weather units to download historical information through telephone lines or directly to a laptop computer. Many growers find these weather stations invaluable in their day-to-day operations, especially in the area of disease prevention.

CHAPTER 24

Vineyard and Winery Relations

F OR GROWERS, choosing a winery to supply with grapes and developing a sound relationship with the winery owners or managers is perhaps one of the most critical elements of making a commercial vineyard a financial success. It is important that grower and winemaker share the same basic goals and objectives, get to know each other, and be able to view the overall situation from each other's perspective. It has long been my assertion that, during industry educational events with breakout sessions for grape growing and winemaking, the grape growers should sit in on the winemaking lectures and winemakers should attend viticulture seminars.

To maintain a healthy relationship with a winery, the grower needs to be able to predict crop size on each block within a certain margin of error because the winemaker needs to know how much of a crop to expect at the crush pad so there is enough labor to process the fruit and sufficient tank space to store the fruit once it is pressed. Top-notch growers can accurately predict crop size within 10 percent, year in and year out. Contracts between grower and winemaker should be negotiated in good faith and renegotiated when conditions dictate. Good relationships between growers and winemakers can and do exist, but there will always be seasons that try the patience of both.

What Makes Good Wine?

Wine quality is driven by adaptation of a variety to the area, having an appropriate crop for the size and condition of the vines, superior cultural practices, sound insect and disease control, and, perhaps most agonizingly, the weather. Crop size does indeed affect wine quality, but crop size varies considerably by variety, and there is a point at which further

As long as fruit remains sound and weather conditions allow, fruit can hang to reach the thresholds a winemaker is looking to achieve. When clusters begin to fall apart and a high percentage of berries have shriveled, it's time for everyone to realize that things are not going to get any better.

diminution of a crop will not result in any improvement in fruit quality. For many varieties, an average of four to five tons per acre, if nothing else is limiting, can produce exceptional quality fruit. For varieties like 'Chenin Blanc', crops of up to ten tons per acre can yield the same quality as vineyards bearing half that. This figure, however, is the exception and certainly not the rule. Some of the very best winemakers make harvest decisions solely based on the flavor of the berries and not on a numerical assessment of soluble solids or acids. Most use a combination of the two. Maintaining moderately vigorous vines, with ample water, nutrients, and canopy management, will make the most of any growing season. This delicate balance can be easily upset by one missed spray that results in insect infestation or a disease outbreak that can affect fruit and, ultimately, wine quality. Then, of course, there is the weather.

Overabundant rain between véraison and harvest can cause a decrease in accumulated soluble solids even while acids continue to fall and pH levels rise. If clear weather is in the forecast, it's probably best in most situations to ride out the rain and let the fruit once again approach desirable sugar levels. Tropical rain events can cause excessive berry splitting and bring on a plethora of fungal and bacterial pathogens that degrade fruit quality very quickly. Although the High Plains is less prone to high rainfall events near harvest, there have been years when those growers have faced rapidly deteriorating conditions that negatively affected both yields and fruit quality. This is farming, and farming is a gamble, but good gamblers also know when to cut their losses.

Growing for Your Market

Not every bottle of wine from every winery is intended to be sold for the high-end market. There is also a good market for moderately priced wines. For those wines, higher yields can produce sound fruit of acceptable commercial quality. Such fruit is usually sold for 20 to 30 percent less than superior fruit, but growers can remain profitable by increasing both vine size and yield. Respectable soluble solids levels can be achieved with these increased yields, but there is usually a reduction in the levels of flavor components needed for the best wine. It is imperative that those distinctions be made up front between grower and buyer to avoid a misunderstanding at or after harvest. Contracts should clearly specify desired maturity parameters and any crop size limitations.

Recognizing the Inevitable

There are years and conditions that produce less than optimal results. The most common problem is rain, but when other conditions slow or halt the ripening of fruit, a decision needs to be made on when to pick. After a significant amount of berry shriveling has occurred, when fruit splitting or rot begins, or once the rachis or other parts of the cluster stem start to collapse, berry quality is only going to get worse, not better.

Winemakers who want to avoid pre-harvest fungicide applications to fruit that is beginning to rot need to assume some of the responsibility for a deteriorating situation and not be overly dogmatic. At some point you simply need to get the fruit out of the field. If the grower has done due diligence, wineries should agree to accept the fruit, perhaps at a reduced or even greatly reduced price. If growers have been negligent or have misrepresented conditions in the vineyard, then they should not be surprised when the fruit is simply rejected. This situation should never occur. Winemakers should visit vineyards to become familiar with the crop load and conditions of acreage they have contracted. Communications should be continuous, and representative fruit sampling should be an ongoing activity.

Shared Risk

The Texas wine industry will grow only when both growers and wineries prosper. For growth and prosperity in the industry, there has to be shared risk. Growers simply cannot expect winemakers to constantly pull rabbits out of their hats and make acceptable wine from inferior fruit. Likewise, winemakers need to recognize that there are just some years that appear to be born from hell. Stay in the industry long enough, and you remember those vintages, the ones that most would rather forget. When growers consistently try the patience of wineries, they should not be surprised when contracts get canceled. Likewise, when wineries throw growers under the bus because of circumstances that were beyond their control, their fruit will most likely go elsewhere in years to come. For this industry to prosper, there must be not only common goals but also shared risk.

For growers and wineries to prosper together, and for a true Texas wine industry to flourish, there must be mutual understanding, cooperation, and a sharing of risk.

Glossary

airblast sprayer. A tractor-powered or tractor-pulled sprayer used to apply materials, typically fungicides or insecticides, to the canopy of a vineyard.

air drainage. The flow of colder air away from a vineyard site, usually as a function of relative elevation or slope.

anion. A negatively charged ion that is not bound chemically by soil exchange sites and is consequently subject to movement via water.

banding. The application of nutrients in a strip next to the trellis to saturate a given area with a high concentration of that nutrient.

Brix. Fermentable soluble solids, measured in degrees. While soluble solids are primarily sugars, they also consist of some proteins and amino acids.

broadcasting. The application of nutrients across the majority of a vineyard floor.

bud. The compressed precursor of a green shoot, which is borne at the axil of a cane.

budbreak. The beginning of growth in the spring whereby green tissue emerges from previously dormant buds.

bull canes. Canes exceeding seven-eighths of an inch in diameter that are excessively vegetative and typically unfruitful and cold tender.

calyptra. The green tissue covering the anthers and ovary of a grape flower prior to bloom.

cane. The woody annual growth of a vine that results from the maturation of a green shoot.

carbohydrates. Compounds in plants created through photosynthesis that then act as the source of energy for plant functions.

cation. A positively charged ion. Positively charged ions are attracted and held by soil particles that are negatively charged.

cation exchange capacity (CEC). The measure of how well a soil can hold positively charged ions so that they remain available for uptake by plants.

chelate. A form of fertilizer that protects specific cations from being rendered unavailable by soil pH or chemistry.

climacteric. Signifying a period of increased respiration in a fruit that has the ability to further ripen after harvest. A nonclimacteric fruit will not show any increase in sugar content after harvest.

cordon. Woody grapevine tissue in the form of lateral extensions of the trunk, usually trained on a support wire in the vineyard. Cordons bear the spurs on which retained nodes, left after dormant pruning, give rise to a crop the next season.

cultivar. A coinage from the phrase "cultivated variety," which refers to the name given to a commercially grown variety.

cuticle. The waxy surface of a leaf.

cytokinin. A class of plant growth regulators responsible for cell division in roots and shoots.

devigorate. An act, such as overcropping or summer pruning, that reduces the vegetative vigor of a vine.

dormancy. The period of winter rest associated with temperate perennial plants.

emitter. Part of an irrigation system, usually attached to a lateral polyethylene pipe designed to deliver a specific flow of water.

Geneva Double Curtain. A divided canopy training system developed by Nelson Shaulis to efficiently manage highly vigorous vines by increasing light exposure to the foliage.

hardiness. The ability of a plant to withstand subfreezing temperatures.

instar. A developmental stage of an organism, such as that of an insect in its progression toward maturity.

lag phase. A period of grape berry development in which visible berry size increase is slowed but flavor components and seeds continue to progress.

lignification. The process in which green plant tissues become woody as a result of lignin deposition in those tissues.

nematode. Any roundworm of the phylum Nematoda. A significant number of the thousands of nematode species are pathogenic to grapevines and other plants.

node. The point on a shoot or cane where a leaf is or was attached. In grapevines, there are typically three buds located at each node.

overcrop. Retaining an excessive amount of fruit on a vine relative to optimal capacity.

own-rooted. A vine that is propagated without the use of rootstock.

pathovar. A coinage from the term "pathogen variety" signifying a specific strain of a pathogen. For example, the pathogen *Xylella fastidiosa* infects a wide variety of plants, but it is the specific strain *Xylella fastidiosa fastidiosa* that is the pathovar responsible for Pierce's disease in grapevines.

periderm. Woody tissue that forms on the surface of green annual shoots as a part of the cane maturation process.

petiole. The leaf stem. Petioles attach the leaf blade to a shoot, and visual observation of petiole condition is used to determine the nutritional needs of grapevines.

pH. A logarithmic measurement of the relative acidity or alkalinity of a soil or a solution. The pH scale runs from 1 to 14, with a pH of 7 being neutral and signifying equal numbers of hydroxyl and hydrogen ions dissolved in solution.

phenology. The study and description of growth phases of a plant.

phenoxy herbicide. Any one of a group of herbicides that mimic plant growth regulators and that, when sprayed on plants, cause rapid, unregulated growth, eventually killing the plant. The most problematic of these compounds for grape growers is 2,4-Dichlorophenoxyacetic acid because of the volatile nature of this herbicide and the tremendous sensitivity of grapevines to injury from it.

pheromone. A chemical component that triggers a behavioral response from an individual. In agriculture, pheromones are used to attract or confuse insects in order to monitor or control their numbers.

phloem. Conductive tissue in plants responsible for the transport of organic compounds such as those produced through photosynthesis.

phylloxera. A very small, pale yellow insect that feeds on the roots and foliage of grapevines.

Pierce's disease. A lethal disease of susceptible species and cultivars of grapevines caused by the bacterium *Xylella fastidiosa*. The disease is named after M. B. Pierce, who first identified the disease near present-day Anaheim, California.

procumbent. Growth that trends toward the ground as opposed to upright growth that proceeds away from the earth.

pruning weight. The weight of one-year-old dormant wood removed during winter pruning. Pruning weight is used to measure relative vigor of vines.

PTO. Power take-off, or the rearward facing shaft on a tractor. Powered by the engine, it is used to drive mowers, sprayers, or other farm implements.

rachis. The main axis or stemlike tissue of a grape cluster. The rachis is attached to the shoot of a grapevine and bears flowers and then berries.

rootstock. A specific species or variety of grapevine upon which the cultivated variety is grafted to create a single vine. Rootstocks are utilized to overcome chemical or biological limitations of a given vineyard site or to improve vine vigor.

rouging. The act of removing a vine from a vineyard. This may be done to prevent a disease from spreading to other grapevines.

SAR. Sodium absorption ratio, or the measurement of the amount of sodium in soil or water in comparison to calcium and magnesium.

scion. The grapevine variety associated with the aboveground portion of the vine responsible for bearing fruit.

shatter. The phenological stage of growth in which unfertilized berries fall from the flower cluster.

shoot. The green annual growth of a grapevine that bears foliage and where fruit may or may not be present.

shoot positioning. The manipulation of the annual growth of a vine to enhance sunlight exposure of foliage or fruit.

sink. The destination for transported photosynthates produced by photosynthesis and supporting plant growth or fruit maturation.

sporulation. Term usually associated with the production and dissemination of spores by a pathogen.

spurs. Woody, semiperennial, vertical extensions of cordons that bear fruiting canes.

stomate (*pl.,* stomata). Minute epidermal pores on foliage through which gas exchange and respiration in leaves occur.

training system. A systemized approach to the spatial arrangement of permanent and annual growth of a perennial plant.

trellis. A structure typically consisting of posts and wire used to support the growth of grapevines.

vector. In agriculture, the biotic agent, such as an insect, responsible for the transmission of a specific disease.

véraison. The phase of fruit ripening in which 10 percent of fruit begin to change color or change from opaque to translucent. Véraison is typically associated with about 8 percent soluble solids.

vinifera (*Vitis vinifera*). A species of grape native to an area extending from south-central Europe and the Caucasus Mountains region into Turkey. Varieties that are pure or have a high percentage of *vinifera* lineage are considered to have the highest wine quality.

xylem. The water-conducting vessels within plants.

Index

Pages with photos are indicated in **bold** type.

2,4,5-T, 196

2,4-D, 13, 196–97. *See also* phenoxy herbicide

ABA. *See* abscisic acid

abscisic acid, 87, 92

acclimation, 32, 62–64

acidic: soils, 15, 17, 45

acidity: and nutrient availability, 158, 165; and cotton root rot, 27; fruit, 77–79, 105, 237. *See also* pH

Agrobacterium vitis. *See* crown gall

air drainage, 15, 48–50

airblast sprayer, 4, 182, 225, 227–229

alkaline: soils, 15, 17, 29, 45; and cotton root rot, 27

alkalinity and nutrient availability, 158, 165–66

alluvial, 15, 27, 66

Alternaria. *See* sour rot

aluminum toxicity, 17, 165

Alvarinho, 114

annual ryegrass, 199–200, 231

anthracnose, 168–69

apical dominance, 133, 159

Armellaria mellea. *See* post oak root rot

Aspergillis. *See* sour rot

Bacillus thuringiensis. *See* BT

bearing capacity: 124; prediction, 124

beneficial fungi, (cotton root rot), 29

beneficial organisms, 29

berry growth, 76–78

bird netting, 223

birds eye rot. *See* Anthracnose

bitter rot. *See* summer fruit rot complex

black measles, 186

black rot: 168–71; conditions for infection table, 171

'Black Spanish', 17, 37, 106–07, 137, 189

'Blanc DuBois', 17, 106, 137, 168, 189

Bordeaux mixture (fungicide), 168, 173

Bordeaux varieties, 107–10

boron: nutrient, 81, 150; deficiency, 76, 160–61; petiole analysis values, 163–165; in water, 207

bot canker. *See* canker (fungal)

Botryosphaeria dothidea: *See* summer fruit rot complex. *See also* canker (fungal)

Botrytis, 148, 177–79

brix, 51, 79, 169

BT, 218

bud(s): axillary, 74; compound, 70; dormancy,87–88; and frost,31; initiation,74, 147; injury (general), 33, 171, 90; injury (due to delayed pruning), 137; force, 59; formation, 83; when pruning, 62, 124–26, 141; fruitfulness, 68, 70, 74, 83, 124, 127–28, 141, 143, 148, 152; primary, 70–71; secondary, 70–71; tertiary, 71

budbreak: delayed, 208; general, 74, 78, 88, 126; lateral, 159–160; effects of pruning on, 126–27; uniformity, 118

bunch rot, 69, 178

'Cabernet franc', 110

'Cabernet Sauvignon', cluster size, 134; disease, 180; vigor, 66; varietal description, 109

Cabernet Sauvignon, wine, 11

calcium: nutrient, 150, 156, 166; petiole analysis value, 165; in soils, 154 calcium bicarbonate, 29, 97, 166

cane, lesion, 168–69

cane pruning, 118, 137

cane training, 118

canker, fungal, 117–18, 134, 186–88

canopy: air movement, 68; divided, 118–122, 134, 138; hedging, 78; light, 68; management, 70, 124, 128, 139–149; spraying, 182, 227–29. *See* leaf removal

Captan, 180

carbohydrate: depletion, 86; fall accumulation, 183; loading, 86, 90; movement, 200; production, 82, 84, 86, 87; storage, 60, 78, 80, 85, 91, 192; and winter injury, 90

carbon, 150–151

'Catawba,' 118

cation exchange capacity (CEC), 47, 166

certified plant material, 54

'Champanel': varietal, 9, 17; rootstock, 29, 100

'Chardonnay': varietal, 108, 113; wine, 11

chilling requirements, 49, 88

chlorine: as nutrient,150; in water, 207

Cladosporium. See sour rot

climbing cutworms, 215

cluster: collapse, 177, 180, 238; flower, 72–74, 174; injury, 180; loose, 76, 107, 113; loss, 174, 216; as sink, 92; tight, 69, 76, 108, 148, 178

Colletotricum spp. See summer fruit rot complex

'Concord,' 40, 45, 96, 118, 154, 165

copper: as fungicide, 168, 171, 177, 181; as nutrient, 150, 165

Cordon: bilateral, 115, **116**; high-wire, 117; injury, 5, **33**, 90, 117; mid-wire, 117; replacing, 117; training, 115; unilateral, **117**

Cotinus nitida. See green June beetles

Cotton Root Rot: control, 29–30; described, 8, 9, 26–28; diagnosis 28 ; range 27–28; risk, 15, 17, 28, 96; resistance, 99, 101

cover crop, 47, 199–201, 231–232

cross timbers, 14, 21

crown gall, 54, 184–186, 189

CRR. *See* cotton root rot

Cuerna costalis, **24**

Cutting(s), 7, 22, 96

'Cynthiana,' 37

cytokinins, 92, 192

dagger nematodes, 37–38, 100, 190

deer, 62, 99, 221–222

defoliation, 171, 177, 209,215

dicamba, 196

Diplodia corticola. See canker, fungal

Dormancy, 87–88, 92, 98, 108, 113, 133, 160

double trunk, 129, 187

downy mildew, 39, 41–42, 85, 106, 113, 168, 171–174, 180, 183

drought, 5, 15, 21, 28, 39, 42, 43, 90, 146

drought stress, 51, 90, 91, 99, 161, 164, 203, 208

East Texas, 17–18, 22, 98, 168

EC. *See* electrical conductivity

electrical conductivity, 50

Elsinoe ampelina. See Anthracnose

emitter, placement, 204

Erysiphe necator. See powdery mildew

Esc. *See* black measles

eutypa, 118, 186

fanleaf virus, 190

'Favorite,' 17, 107

fertilization, 78, 149–166

floor management, 21, 24, 25, 191–201, 231

floral primordial, 128

fluazifop-p-butyl. See Fusilade®

flutriafol, 29

Foundation Plant Service, 29

fox grape, 39–40

freeze: injury, 5, 14, 32–**33, 63**, 70, 88, 90, 96, 129, **140, 184**, 186; spring, 32; winter, 12–13, 15, 32–33, 80, 87

French-American hybrid, 9, 117, 189

frost: early/fall, 5, 32, 88; protection, 201, 207 ; spring/late, 9, 15, 31–32, 48, 49, 98, 126–127, 154, 201

fruit: abortion, 92; quality, 18, 29, 51, 67, 79, 104, 124, 138, 139, 143, 147–48, 208–09; ripening, 31, 78, 86, 92, 141, 238, ; set, 74–76, 85, 92, 124, 156, 159, 160–61, 194

fungal disease, management, 167

Fusilade®, 196

GBM. *See* grape berry moth

GDC. *See* Geneva double curtain

Geneva double curtain, 118, **121**, 137

'Gewurztraminer': varietal, 111; wine, 11

glassy winged sharpshooter, 211

glufosinate, 196

glyphosate, 55, 196–97, 200–201, 230

goblet, 121

grafting, 93, **94**

grape berry moth, 212–**13**

grape cane borer, 217–**18**

grape cane gallmaker, 219

grape cane girdler, 219

grape tumid gallmaker, 219

grapevine epimensis, 219–**20**

grass, management in the vineyard, 55, 196

green graft vine, 59–60

green june beetles, 216–**17**

Greeneria uvicola. See summer fruit rot complex

'Grenache', varietal, 111

grow tubes, 62–64, 129

growth regulators, 88, 92, 159, 192

Guignardia bidwellii. See black rot

Gulf Coast, 21, 32

GWSS. *See* glass winged sharpshooter

Hail, 5, 9, 13, 33–34, 180

Head pruning. *See* cane pruning

Hedging, 78, 145–47

'Herbemont', varietal, 8

herbicide, application 194–97, 230 ; calculation; calibration/sprayer usage check list, 198–99; injury, 13, 55, 62, 196–**97**, 201; post-emergent, 195–96, 230, 199; pre-emergent, 195, 201; sprayer, 229–30

High Plains, 11, 12–13, 22, 33, 50, 80, 109, 110, 158, 168, 176, 204, 215, 222, 237

Hill Country, 11, 15, 17, 21, 28, 30, 33, 43, 49–50, 90, 98, 102, 108, 158, 180

Hormones, 192, **194**

hybrid, 9, 37, 40, 98, 99, 105, 129, 165, 210. *See also* French-American hybrids.

insecticide, 167, 189, 196, 219

internode, length, 118, 125, 134, 152, 208

iron: 17, 45, 85, 150, 158–59, target value, 165

Isariopsis leaf spot. *See* leaf blight.

'Jacquez', varietal, 106

labor, 4–5, 16, 58, 83, 115, 118, 127, 138, 143, 147, 235

Landers, Andrew, 182

Lasiodiplodia theobromae. See canker, fungal.

leaf, removal, 147–48, 178

leaf roll virus, 188–89

leafhopper, 211, 214–15

'Lenoir', varietal, 8, 106

Lipe, Bill, **10**

Lower Rio Grande Valley, 16

Lyda, Stuart, 27–28

Macrophoma rot. *See* summer fruit rot complex.

magnesium,: 85, 149, 150, 154, 156–**57**, 163, 166; target values, 165

'Malbec', varietal, 109

'Mataro', varietal, 111

McEachern, George Ray, **10**

Meloidogyne incognita. See root knot nematode.

'Merlot', varietal, 108–109, 134; wine, 11

mites, 220

Mortensen, Ernest, 9

Mortensen, John, 106
'Mourvedre,' varietal, 111
mower, 226–27
mulch, 47, 215–16
Munson, Thomas Volney, 8–9, 93, 99
Muscadinia, 35–37
Mustang grape, 40–41

'Nebbiolo,' varietal, 114
Nematode: dagger, 37, 38, 100, 190; root
     knot, 29, 37, 41, 99, 100; variouse, 46,
     96, 98, 99
Neofusicoccum parvum. See canker, fungal.
neonicotenoid, 21, 24, 189
'Nero d'Avola,' varietal, 114
'Niagara,' 45, 96, 118
Nitrogen: 66, 78, 81, 85, 92, 125, 149–154,
     156, 163, 203; target values, 165
Norflurazon. *See* Solicam®
North Texas, 11, 33
'Norton,' 37, 118

Onderdonk, Gilbert, 7–8
open lyre, 121–**22,** 137
organic approach, 167, **170**–71, 181, 218,
     219
organic matter, 47–48, 55, 200, 216,
     231–32
overcropping, 86, 141, 161
own-rooted, 17, 96, 210

Paralobesia viteana. See grape berry moth.
paraquat, 196
PD, *See* Pierce's disease
Penicillium. See sour rot.
Perry, Ron, 9, 19–20
petiole: analysis/testing/sampling, 46, 85,
     152, 162–64, 166; values 92,161, 164;
     values table, 165
'Petit Mansang,' 113
'Petit Verdot,' 114
pH: and nutrient availability, 45; soil,
     44–45
phenoxy herbicides, 13, 196
phloem-feeder. *See* leafhopper

Phomopsis, 118, 174–**75, 76,** 182, 187
phosphorus: 149–50, 157–58, 166; target
     value, 165
Phylloxera, 8, 93, 96, 98–100, 102, 210–
     211
Phymatotrichopsis omnivore. See cotton
     root rot
Pierce's disease: described, 19; diagnosis,
     25; history, 8–9; management, 23–**26,**
     105, 211; range, 14, 19–20; risk, 12, 17,
     21–23; symptoms, 25
Pinot Noir, wine, 11
Plasmopara viticola. See downy mildew.
Plume moth, 219
Poast®, 196
post oak root rot, 96
potassium: 92, 149, 150, 154, 156–57, 163,
     166; target value, 165
powdery mildew, 12, 37, 106, 107, 168,
     176–**78**
pruning: balanced, 124–25; basic, 129–38;
     during active growth, 78; delayed, 126–
     28, 141; dormant, 70, 118, 123; double,
     31, 127–28; rough, 127; weight, 124–25

Qualia, Frank, 8–9

Randolph, Uriel, 9
red blotch, 189
red leaf. *See* leaf roll virus.
Red River Valley, 14–15
Rhizopus. See sour rot.
Rhône Varieties, 110–111
Riesling, 11
ripe rot. *See* summer fruit rot complex.
root: anchor, 91, 210; feeder, 158, 192–**93,**
     231; growth, 31 47–48, 59, 81, 91–92,
     129, 161, 192, 194, 202; hormone pro-
     duction, 92, 192; system, 61–62 , 91,
     129, 191; tips, 91–92, 152
rootstocks: 1103P, 29; 5BB, 29; 5C, 29,
     101; Champanel, 29, 100; Dogridge, 29,
     100; St. George, 42; table of common,
     100
rouging, 21, 23, 26, 190

Sainte Genevieve, 13
'Sangiovese,'varietal, 113, 134, 141
SAR: *See* sodium absorption ratio. *See* also
    systemically acquired resistance.
Sauvignon Blanc, wine, 11, 106
section 18 label, 30
self-rooted. *See* own-rooted.
'Sémillon,' varietal, 108
senescence, 85, 87, 164
Septoria ampelina. See Septoria leaf spot.
Septoria Leaf Spot, **181**
Sethoxydim. *See* Poast®
sharpshooter, 25, 26, 201, 211, **212**, 214
Shaulis, Nelson, 11, 118, 124, 164
'Shiraz,' varietal, 110
shoot: positioning, 83, **119** 122, 142–43;
    mechanical, 143, **145**; thinning,
    139–141
site: preparation, 54; selection, 3, 5, 15, 21,
    23, 26, 44–52, 53
sodium: 15, 27, 50, 97, 207; target values,
    165
sodium absorption ratio, 15, 50, 207
Soil: acidic, 15, 17, 27, 39, 45, 98, 165; al-
    kaline, 15, 17, 27, 29, 45, 101, 158, 165–
    66; alluvial, 15, 27, 39, 66; clay; clay, 17,
    47–48, 60, 101, 203; compaction, **195**,
    201, 203, 226 15; depth, 15, 50, 67, 161,
    166; drainage, 16, 47–48, 52, 161; sam-
    pling, 44, 46; structure, 47–48
Solicam®, 195
sour rot, 179
spacing: row, 65–66, 118, 121; vine, 60,
    66–68, 121
Spotts, Bob, 170
sprinklers, 32, 203–204, 206–207
sterol-inhibitor, 180
strobilurin, 180
suckering, 29, 101
sulfur: fungicide, 168–69, 171, 176–177,
    181; nutrient, 150; soil amendment, 2
summer grape, 37
summer rot, 148, 178–80
'Syrah,' varietal, 110–11
systemically acquired resistance, 182

table grapes, 16
'Tannat,' varietal, 114
'Tempranillo,' varietal, 112
tilling, 191–92
'Tinto Cao,' varietal, 114
'Touriga Nacional,' varietal, 114
tractor, 4, 65, 68, 69, 194, 197–98, 223,
    225–30, 233
Trans-Pecos, 13–14, 222
Trellis: construction/maintenance, 2,
    57–59, 127; system, 138
triazole, 29

Val Verde Winery, 8–9
Véraison, 31, 78–79, 84, 145, 156, 162,
    164–65, 168, 179, 208–209, 223,
    237
'Vermentino,' varietal, 113
'Victoria Red,' varietal, 107
'Viognier,' varietal, 111
Vitis acerifolia, 102
Vitis aestivalis, 37, 106
Vitis amurensis, 43
Vitis arizonica: galvinii, 38; glabra, 38
Vitis berlandieri, 29, **30**, 38, 39, 96, 99,
    100, 158
vitis bourquiniana, 106
Vitis candicans, 40, 99, 100, 102
Vitis caribaea, 43
vitis champini, 99, 100
Vitis cinerea, 39
Vitis doaniana, 102
Vitis labrusca, 8, 37, 39–40, 45, 88, 96,
    118
Vitis lincecumii, 37
Vitis monticola, 43, 102
Vitis munsoniana, 36–37
Vitis mustangensis, 40–41, 99
Vitis popenoii, 37
Vitis riparia, 41, 98, 100
Vitis rotundifolia, 36–37, 102
vitis rufotomentosa, 102
Vitis rupestris, 42, 98–100
Vitis shuttleworthii, 40
Vitis vinifera, 43

Walker, Andy, 29, 105

water: deficit, 84; quality, 15, 50, 207; requirements, 50, 231; salinity, 50–51

weed, management, 2, 4, 13, 55, 62, 91,
191, 195–200

well capacity, 50, 207

winter grape, 38–39

winter injury, 9, 14, 32–33, 64, 82, **89**–90,
117, 124, 125, 126, 134, 139, 153

Xiphinema index. See dagger nematode.

Xylella fastidiosa, 19, 22

yield, 51, 55, 65–67, 70, 80, 86, 115, 117,
118, 121–22, 138, 143, 161, 192, 212,
237

zinc, 45, 76, 85, 150, 158–60, 163, 165–66,
200